Kristen Nehemiah Horst (Ed.)

# Malva parviflora

**Kristen Nehemiah Horst (Ed.)**

# Malva parviflora

## North Africa, Malva, Malvaceae, Malvales, Rosids

**Dign Press**

# Contents

# Malva parviflora

| Malva parviflora | |
|---|---|
| 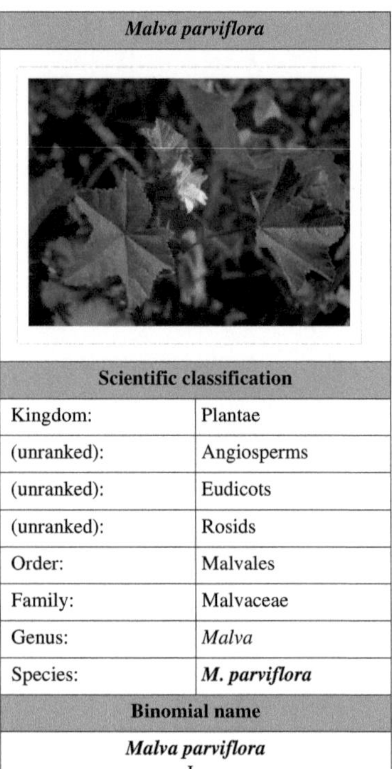 | |
| **Scientific classification** | |
| Kingdom: | Plantae |
| (unranked): | Angiosperms |
| (unranked): | Eudicots |
| (unranked): | Rosids |
| Order: | Malvales |
| Family: | Malvaceae |
| Genus: | *Malva* |
| Species: | ***M. parviflora*** |
| **Binomial name** | |
| *Malva parviflora*<br>L. | |

***Malva parviflora*** is an annual or perennial herb that is native to Northern Africa, Europe and Asia and is widely naturalised elsewhere.[1] Common names include **cheeseweed**,[1] **cheeseweed mallow**, **Egyptian mallow**,[1] **least mallow**, **little mallow**,[1] **mallow**,[2] **marshmallow**,[2] **small-flowered mallow**,[3] **small-flowered marshmallow**[4] and **smallflower mallow**.[2] *M. parviflora* leaf extracts possess anti-inflammatory and antioxidant activities.[5] It has a decumbent or erect habit, growing to 50 cm in height.[3] The broad leaves have 5 to 7 lobes and are 8 to 10 cm in diameter.[3] It has small white or pink flowers with 4 to 6 mm long petals.[3]

## References

[1]  "*Malva parviflora*" (http://www.ars-grin.gov/cgi-bin/npgs/html/taxon.pl?90031). *Germplasm Resources Information Network (GRIN)*. United States Department of Agriculture, Agricultural Research Service, Beltsville Area. . Retrieved 2008-06-02.

[2]  "*Malva parviflora*" (http://florabase.dec.wa.gov.au/browse/profile/4961). *FloraBase*. Department of Environment and Conservation, Government of Western Australia. .

[3]  New South Wales Flora Online: *Malva parviflora* (http://plantnet.rbgsyd.nsw.gov.au/cgi-bin/NSWfl.pl?page=nswfl&lvl=sp& name=Malva~parviflora) Royal Botanic Gardens & Domain Trust, Sydney, Australia

[4]  "*Malva parviflora* L." (http://www.flora.sa.gov.au/cgi-bin/texhtml.cgi?form=speciesfacts&family=Malvaceae&genus=Malva& species=parviflora). *Electronic Flora of South Australia Fact Sheet*. State Herbarium of South Australia. .

[5]  Bouriche H, Meziti H, Senator A, Arnhold J"Anti-inflammatory, free radical-scavenging, and metal-chelating activities of Malva parviflora." *Pharm Biol.* 2011 May 19;

## External links

- *"Malva parviflora"* (http://fm1.fieldmuseum.org/vrrc/?page=view&id=30110). Neotropical Herbarium Specimens. Retrieved 2008-06-20.
- Jepson Manual Treatment (http://ucjeps.berkeley.edu/cgi-bin/get_JM_treatment.pl?5042,5084,5087)
- USDA Plants Profile (http://plants.usda.gov/java/profile?symbol=MAPA5)
- Photo gallery (http://calphotos.berkeley.edu/cgi/img_query?query_src=photos_index&where-taxon=Malva+parviflora)

# North Africa

**North Africa** or **Northern Africa** is the northernmost region of the African continent, linked by the Sahara to Sub-Saharan Africa. Geopolitical, the United Nations definition of Northern Africa includes eight countries or territories; Algeria, Egypt, Libya, Morocco, South Sudan, Sudan, Tunisia, and Western Sahara.[1] Algeria, Morocco, Tunisia, Mauritania, and Libya together are also referred to as the Maghreb or **Maghrib**, while Egypt and Sudan are referred to as **Nile Valley**, Egypt is a transcontinental country by virtue of the Sinai Peninsula, which is in Asia. North Africa also includes a number of Spanish *possessions*, Ceuta and Melilla (tiny Spanish exclaves or islets off the coast of Morocco). The Canary Islands and the Portuguese Madeira Islands, in the North Atlantic Ocean northwest of the African mainland, are sometimes included in considerations of the region.

Northern Africa (United Nations geoschemeUN subregion) geographic, including aboveClockwise from the left:  Western Sahara, Morocco, Algeria, Tunisia, Libya, Egypt, Sudan, South Sudancontinued from right to left:  Chad, Niger, Mali, Mauritania

The distinction between Northern Africa and the rest of Africa is historically and ecologically significant because of the effective barrier created by the Sahara. Throughout history this barrier has culturally separated the North from the rest of Africa and, as the seafaring civilizations of the Phoenicians, Greeks, Romans and others facilitated communication and migration across the Mediterranean, the cultures of North Africa became much more closely tied to Southwestern Asia and Europe than Sub-Saharan Africa. The Islamic influence in the area is significant, and North Africa, along with the Middle East, is a major part of the Arab World.

Some researchers have postulated that North Africa, and not East or South Africa, was the original home of the modern humans who first trekked out of the continent.[2][3][4]

## Geography

The Atlas Mountains, which extend across much of Morocco, northern Algeria and Tunisia, are part of the fold mountain system which also runs through much of Southern Europe. They recede to the south and east, becoming a steppe landscape before meeting the Sahara desert which covers more than 75% of the region. The sediments of the Sahara overlie an ancient plateau of crystalline rock, some of which is more than four billion years old.

Sheltered valleys in the Atlas Mountains, the Nile valley and delta, and the Mediterranean coast are the main sources of good farming land. A wide variety of valuable crops including cereals, rice and cotton, and woods such as cedar and cork, are grown. Typical mediterranean crops such as olives, figs, dates and citrus fruits also thrive in these areas. The Nile valley is particularly fertile, and most of population in Egypt and Sudan lives close to the river. Elsewhere, irrigation is essential to improve crop yields on the desert margins.

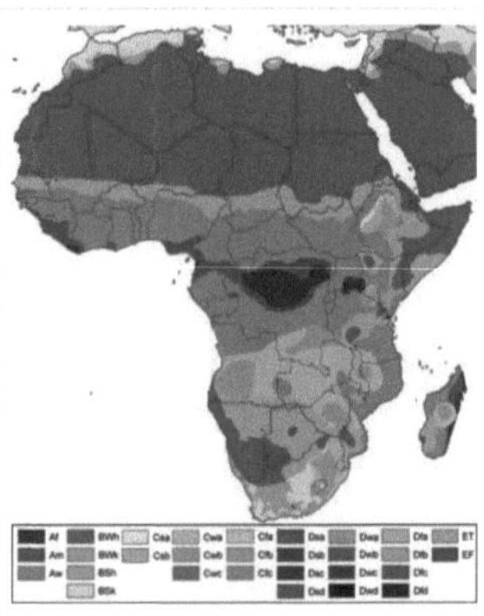

North Africa, consisting of the Sahara and north, in the northern red climatic zone and northwards

## Territories and regions

| Countries and territories | Area (km²) | Population | Density (per km²) | Capital | GDP (Total) | GDP per capita | Currency | Government | Official languages |
|---|---|---|---|---|---|---|---|---|---|
| Algeria | 2,381,741 | [5]34,994,937 | 14.5 | Algiers | $254.7 billion (2010 est.)[5] | $7,400 (2010 est.) | Algerian dinar | Presidential republic | Arabic (official), Berber (national) |
| Egypt | 1,001,450 | 80,471,869 | 80.4 | Cairo | $500.9 billion (2010) | $6,200 (2010) | Egyptian pound | Semi-presidential republic | Arabic |
| Libya | 1,759,540 | 6,461,454 | 3.7 | Tripoli | $89.03 billion (2010)[6] | $13,800 (2010) | Libyan dinar | Provisional authority | Arabic (official), Berber (national) |
| Morocco | 710,850 | [7]32,226,056 | 70.8 | Rabat | $153.8 billion (2010)[8] | $4,900 (2010) | Moroccan dirham | Constitutional monarchy | Arabic and Berber (both official) |
| South Sudan | 619,745 | 8,260,490 | 13.3 | Juba | $13.227 billion (2011 estimate)[9] | $1,546 (2011 estimate) | South Sudanese pound | Federal republic | English |
| Sudan | 1,886,068 | 30,894,000 | 16.4 | Khartoum | $98.79 billion (2010) prior to break up[10] | $2,200 (2010)—prior to break-up | Sudanese pound | Federal republic (Authoritarian) | Arabic |

| | | | | | | | | | | |
|---|---|---|---|---|---|---|---|---|---|---|
| Tunisia | 163,610 | 10,589,025 | 64.7 | Tunis | $100.3 billion (2010)[11] | $9,500 (2010) | Tunisian dinar | Republic | Arabic |
| Western Sahara[12] Morocco | 266,000 | [13]320,000 | 1.2 | El Aaiún (Laâyoune) (controlled by Morocco) | $900 million (2007)[14] | $2,500 (2007) | Moroccan dirham | Constitutional monarchy | Arabic and Berber (both official), Hassaniya (recognized) |
| **Total, North Africa** | 7,904,959 | 195,637,341 | 24.7 | | $1.2 trillion | $5,700 | | | |

**Source:**
- The World Factbook, United States Central Intelligence Agency (CIA), 11 February 2011.[15]

## People

The inhabitants of North Africa are generally divided in a manner roughly corresponding to the principal geographic regions of North Africa: the Maghreb, the Nile Valley, and the Sahara. Northwest Africa on the whole is believed to have been inhabited by Berbers since before the beginning of recorded history, while the eastern part of North Africa has been home to the Egyptians and Nubians. Ancient Egyptians record extensive contact in their Western desert with people that appear to have been Berber or proto-Berber.

The official language or one of the official languages in all of the countries in North Africa is Arabic. The largest ethnic groups in North Africa are the Arabs and Berbers. North Africa is predominantly Muslim, with Christian and Jewish minorities.

## Culture

The people of the Maghreb and the Sahara speak various dialects of Berber and Arabic, and almost exclusively follow Islam. The Arabic and Berber groups of languages are distantly related, both being members of the Afro-Asiatic family. The Sahara dialects are notably more conservative than those of coastal cities (see Tuareg languages). Over the years, Berber peoples have been influenced by other cultures with which they came in contact: Nubians, Greeks, Phoenicians, Egyptians, Romans, Vandals, Arabs, and lately Europeans. The cultures of the Maghreb and the Sahara therefore combine indigenous Berber, Arab and elements from neighboring parts of Africa and beyond. In the Sahara, the distinction between sedentary oasis inhabitants and nomadic Bedouin and Tuareg is particularly marked.

The diverse peoples of the Sahara are usually categorized along ethno-linguistic lines. In the Maghreb, where Arab and Berber identities are often integrated, these lines can be blurred. Some Berber-speaking North Africans may identify as "Arab" depending on the social and political circumstances, although substantial numbers of Berbers (or **Imazighen**) have retained a distinct cultural identity which in the 20th century has been expressed as a clear ethnic identification with Berber history and language. Arabic-speaking Northwest Africans, regardless of ethnic background, often identify with Arab history and culture and may share a common vision with other Arabs. This, however, may or may not exclude pride in and identification with Berber and/or other parts of their heritage. Berber political and cultural activists for their part, often referred to as Berberists, may view all Northwest Africans as principally Berber, whether they are primarily Berber- or Arabic-speaking (see also Arabized Berber).

The Nile Valley through northern Sudan traces its origins to the ancient civilizations of Egypt and Kush. The Egyptians over the centuries have shifted their language from Egyptian to modern Egyptian Arabic, while retaining a sense of national identity that has historically set them apart from other people in the region. Most Egyptians are

Sunni Muslim and a significant minority adheres to Coptic Christianity. In Nubia, straddling Egypt and Sudan, a significant population retains the ancient Nubian language but has adopted Islam. The Sudan, which now encompases the modern states of Sudan and South Sudan is home to a largely Arab Muslim population, although there remains significant non-Arab (through Muslim) populations in the north (Nubians), west (Fur, Masalit and Zaghawa) and south (Nuba) of Sudan and largely non-Muslim populations in South Sudan.

North Africa formerly had a large Jewish population, almost all of whom emigrated to France or Israel when the North African nations gained independence. A smaller number went to Canada. Prior to the modern establishment of Israel, there were about 600,000-700,000 Jews in Northern Africa, including both Sfardīm (refugees from France, Spain and Portugal from the Renaissance era) as well as indigenous Mizrāḥîm. Today, less than fifteen thousand remain in the region, almost all in Morocco and Tunisia, and are mostly part of a French-speaking urban elite. (See Jewish exodus from Arab lands.)

# History

## Antiquity and ancient Rome

The most notable nations of antiquity in western North Africa are Carthage and Numidia. The Phoenicians colonized much of North Africa including Carthage and parts of present day Morocco (including Chellah, Mogador and Volubilis[16]). The Carthaginians were of Phoenician origin, with the Roman myth of their origin being that Queen Dido, a Phoenician princess was granted land by a local ruler based on how much land she could cover with a piece of cowhide. She ingeniously devised a method to extend the cowhide to a high proportion, thus gaining a large territory. She was also rejected by the Trojan prince Aeneas according to Virgil, thus creating a historical enmity between Carthage and Rome, as Aeneas would eventually lay the foundations for Rome. The Carthaginians were a commercial power and had a strong navy, but relied on mercenaries for land soldiers. The Carthaginians developed an empire in the Iberian Peninsula and Sicily, the latter being the cause of First Punic War with the Romans.

The first Roman emperor native to North Africa was Septimius Severus, born in Leptis Magna in present-day Libya.

Over a hundred years and more, all Carthaginian territory was eventually conquered by the Romans, resulting in the Carthaginian North African territories becoming the Roman province of Africa in 146 B.C.[17] This led to tension and eventually conflict between Numidia and Rome. The Numidian wars are notable for launching the careers of both Gaius Marius, and Sulla, and stretching the constitutional burden of the Roman republic, as Marius required a professional army, something previously contrary to Roman values to overcome the talented military leader Jugurtha.[18]

North Africa remained a part of the Roman Empire, which produced many notable citizens such as Augustine of Hippo, until incompetent leadership from Roman commanders in the early fifth century allowed the Germanic barbarian tribe, the Vandals, to cross the Strait of Gibraltar, whereupon they overcame the fickle Roman defense. The loss of North Africa is considered a pinnacle point in the fall of the Western Roman Empire as Africa had previously been an important grain province that maintained Roman prosperity despite the barbarian incursions, and the wealth required to create new armies. The issue of regaining North Africa became paramount to the Western Empire, but was frustrated by Vandal victories. The focus of Roman energy had to be on the emerging threat of the Huns. In 468 AD the Romans made one last serious attempt to invade North Africa but were repelled. This perhaps marks the point of terminal decline for the Western Roman Empire. The last Roman emperor was deposed in 476 by the Heruli general Odoacer. Trade routes between Europe and North Africa remained intact until the coming of Islam. Some Berbers were Christians (but evolved their own Donatist doctrine),[19] some were Jewish, and some adhered to their traditional polytheist religion. African pope Victor I served during the reign of Roman emperor Septimius Severus, of Roman/Berber ancestry.[20] The Byzantine reconquest of North Africa from the Vandals began in 533 AD, as Justinian I sent his general Belisarius to reclaim the former Roman province of Africa.

## Arab conquest to modern times

The Arab Islamic conquest reached North Africa in 640 AD. By 670, most of North Africa had fallen to Muslim rule. Indigenous Berbers subsequently started to form their own polities in response in places such as Fez, Morocco, and Sijilimasa. In the eleventh century, a reformist movement made up of members that called themselves Almoravids, expanded south into Sub-Saharan Africa.

The North Africa's populous and flourishing civilization collapsed after exhausting its resources in internal fighting and suffering devastation from the invasion of the Bedouin tribes of Banu Sulaym and Banu Hilal. Ibn Khaldun noted that the lands ravaged by Banu Hilal invaders had become completely arid desert.[22]

The Great Mosque of Kairouan, founded by the Arab general Uqba Ibn Nafi in 670 AD, is the oldest and most important mosque in North Africa;[21] city of Kairouan, Tunisia.

After the Middle Ages the area was loosely under the control of the Ottoman Empire, except Morocco. After the 19th century, the imperial and colonial presence of France, the United Kingdom, Spain and Italy left the entirety of the region under one form of European occupation.

In World War II from 1940 to 1943 the area was the setting for the North African Campaign. During the 1950s and 1960s all of the North African states gained independence. There remains a dispute over Western Sahara between Morocco and the Algerian-backed Polisario Front.

In 2010 - 2011 massive protests swept the region leading to the overthrow of the governments in Tunisia and Egypt, as well as civil war in Libya. Large protests also occurred in Algeria and Morocco to a lesser extent. Many hundreds died in the uprisings.[23]

## Transport and industry

The economies of Algeria and Libya were transformed by the discovery of oil and natural gas reserves in the deserts. Morocco's major exports are phosphates and agricultural produce, and as in Egypt and Tunisia, the tourist industry is essential to the economy. Egypt has the most varied industrial base, importing technology to develop electronics and engineering industries, and maintaining the reputation of its high-quality cotton textiles.

Oil rigs are scattered throughout the deserts of Libya and Algeria. Libyan oil is especially prized because of its low sulphur content, which means it produces much less pollution than other fuel oils.

## Climate change

In 2010, Chad, Niger and Sudan all recorded their hottest all-time temperatures on record. In Chad, the temperature reached 47.6°C (117.7°F) on June 22 in Faya-Largeau, breaking a record set in 1961 at the same location. Niger tied its highest temperature record set in 1998, on also June 22, at 47.1°C (116.78°F) in Bilma. That record was broken the next day, on June 23 when Bilma hit 48.2°C (118.8°F). The hottest temperature recorded in Sudan was reached on June 25, at 49.6°C (121.3°F) in Dongola, breaking a record set in 1987.[24]

## Notes

[1]  According to UN country classification here: http://millenniumindicators.un.org/unsd/methods/m49/m49regin.htm. The disputed territory of Western Sahara (formerly Spanish Sahara) is mostly administered by Morocco; the Polisario Front claims the territory in militating for the establishment of an independent republic, and exercises limited control over rump border territories.

[2]  Was North Africa the Launch Pad for Modern Human Migrations? (http://www.sciencemag.org/content/331/6013/20.summary) Michael Balter, science 7 January 2011: 331 (6013), 20-23. doi:10.1126/science.331.6013.20

[3]  A Revised Root for the Human Y Chromosomal Phylogenetic Tree: The Origin of Patrilineal Diversity in Africa (http://www.cell.com/AJHG/fulltext/S0002-9297(11)00164-9#). Fulvio Cruciani, Beniamino Trombetta, Andrea Massaia, Giovanni Destro-Bisol, Daniele Sellitto, Rosaria Scozzari, The American Journal of Human Genetics - 19 May 2011

[4]  Earliest evidence of modern human life history in North African early Homo sapiens (http://www.pnas.org/content/104/15/6128.full), Tanya M. Smith, Paul Tafforeau, Donald J. Reid, Rainer Grün, Stephen Eggins, Mohamed Boutakiout, Jean-Jacques Hublin, doi:10.1073/pnas.0700747104 PNAS April 10, 2007 vol. 104 no. 15 6128-6133

[5]  "ALGERIA" (https://www.cia.gov/library/publications/the-world-factbook/geos/ag.html). *The World Factbook*. CIA. .

[6]  "LIBYA" (https://www.cia.gov/library/publications/the-world-factbook/geos/ly.html). *The World Factbook*. CIA. .

[7]  "Site institutionnel du Haut-Commissariat au Plan du Royaume du Maroc" (http://www.hcp.ma/). .

[8]  "MOROCCO" (https://www.cia.gov/library/publications/the-world-factbook/geos/mo.html). *The World Factbook*. CIA. .

[9]  "Release of first Gross Domestic Product (GDP) and Gross National Income (GNI) figures for South Sudan by the NBS"" (http://ssnbs.org/storage/GDP Press release_11.08.11.pdf) (PDF). South Sudan National Bureau of Statistics (NBS). 11 August 2011. . Retrieved 2011-09-05.

[10]  "SUDAN" (https://www.cia.gov/library/publications/the-world-factbook/geos/su.html). *The World Factbook*. CIA. .

[11]  "TUNISIA" (https://www.cia.gov/library/publications/the-world-factbook/geos/ts.html). *The World Factbook*. CIA. .

[12]  Under Moroccan administration

[13]  Estimate based on the 2004 Moroccan census. No census specific to the borders of the territory since 1975.

[14]  "WESTERN SAHARA" (https://www.cia.gov/library/publications/the-world-factbook/geos/wi.html). *The World Factbook*. CIA. .

[15]  "The World Factbook" (https://www.cia.gov/library/publications/the-world-factbook/index.html). CIA. . Retrieved 2011-02-11.

[16]  C. Michael Hogan (December 18, 2007). "Volubilis - Ancient Village or Settlement in Morocco" (http://www.megalithic.co.uk/article.php?sid=14906). The Megalithic Portal. . Retrieved 2010-05-23.

[17]  *The Punic Wars 264-146 BC*, by Nigel Bagnall

[18]  Sallust, *De Bello Iugurthino*

[19]  The Berbers (http://www.bbc.co.uk/worldservice/specials/1624_story_of_africa/page66.shtml) BBC World Service: The Story of Africa

[20]  *"Berbers : ... The best known of them were the Roman author Apuleius, the Roman emperor Septimius Severus, and St. Augustine"*, *Encyclopedia Americana*, Scholastic Library Publishing, 2005, v.3, p.569

[21]  Kung, Hans (2006). *Tracing the Way: Spiritual Dimensions of the World Religions* (http://books.google.com/?id=sm0BfUKwct0C&pg=PA248&dq=kairouan+oldest+mosque+north+africa#v=onepage&q=kairouan oldest mosque north africa&f=false). Continuum International Publishing Group. p. 248. ISBN 9780826494238. .

[22]  Populations Crises and Population Cycles (http://www.galtoninstitute.org.uk/Newsletters/GINL9603/PopCrises3.htm), Claire Russell and W.M.S. Russell, Galton Institute, March 1996

[23]  Essa, Azad (February 21, 2011). "In search of an African revolution" (http://english.aljazeera.net/indepth/features/2011/02/201122164254698620.html). Al Jazeera. .

[24]  Masters, Jeff. "NOAA: June 2010 the globe's 4th consecutive warmest month on record" (http://www.wunderground.com/blog/JeffMasters/comment.html?entrynum=1544). *Dr. Jeff Masters' WunderBlog*. Weather Underground. .

## External links

- Human Rights for Indigenous Peoples (http://www.amazighworld.org)
- North Africa's Weather Forecasts and Weather Conditions (http://www.takoumba.com)
- North Africa news and analysis (http://www.north-africa.com)
- Africa Interactive Map (http://www.usaraf.army.mil/MAP_INTERACTIVE/INTERACTIVE_MAP.swf) from the United States Army Africa

# Malva

*Malva was an Dacian settlement in present day Romania, where the Roman city of Romula was built. It is an alternative spelling for the Indian region and state Malwa, the pseudonym of Syrian artist Omar Hamdi, and it is the name of a South African sweet pudding.*

| Malva | |
|---|---|
| 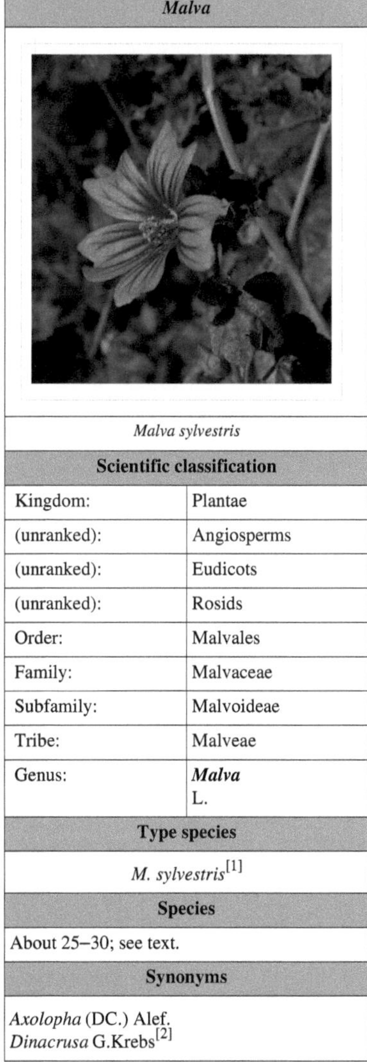 | |
| *Malva sylvestris* | |
| **Scientific classification** | |
| Kingdom: | Plantae |
| (unranked): | Angiosperms |
| (unranked): | Eudicots |
| (unranked): | Rosids |
| Order: | Malvales |
| Family: | Malvaceae |
| Subfamily: | Malvoideae |
| Tribe: | Malveae |
| Genus: | *Malva* L. |
| **Type species** | |
| *M. sylvestris*[1] | |
| **Species** | |
| About 25–30; see text. | |
| **Synonyms** | |
| *Axolopha* (DC.) Alef. *Dinacrusa* G.Krebs[2] | |

**Malva** is a genus of about 25–30 species of herbaceous annual, biennial, and perennial plants in the family Malvaceae (of which it is the type genus), one of several closely related genera in the family to bear the common English name **mallow**. The genus is widespread throughout the temperate, subtropical and tropical regions of Africa,

Asia and Europe.[3] The word "mallow" is derived from Old English "malwe", which was imported from Latin "malva", which originated in Ancient Greek μαλάχη (malakhē) meaning "yellow" or Hebrew מָלוּחַ (malúakh) meaning "salty".[4][5] A number of species, previously considered to belong to Lavatera, have been moved to Malva.

The leaves are alternate, palmately lobed. The flowers are from 0.5–5 cm diameter, with five pink or white petals.

The colour mauve was in 1859 named after the French name for this plant.

## Cultivation and uses

Several species are widely grown as garden flowers, while some are invasive weeds, particularly in the Americas where they are not native.

Many species are edible as leaf vegetables. Known as *ebegümeci* in Turkish, it is used as vegetable in Turkey in various forms such as stuffing the leaves with bulgur or rice or using the boiled leaves as side dish. *Malva verticillata* (Chinese: ; pinyin: *dōngháncài*, Korean:  auk) is grown on a limited commercial scale in China; when made as a herbal infusion, it is used for its colon cleansing properties and as a weight loss supplement.

Very easily grown, short-lived perennials often grown as ornamental plants. Mild tasting young mallow leaves can be a substitute for lettuce, whereas older leaves are better cooked as a leafy green vegetable. The buds and flowers can be used in salads.

Cultivation is by sowing the seeds directly outdoors in early spring. The seed is easy to collect, and they will often spread themselves by seed.

In Catalonia (Southern Europe) they use the leaves to cure stinging nettles sting.

Bodo tribals in Bodoland, Assam (Northeast India) cultivate a sub-species of malva and use it extensively in their traditional cuisine, although its use is not much known among other people of India.

## History

This plant is one of the earliest cited in recorded literature. Horace mentions it in reference to his own diet, which he describes as very simple: "Me pascunt olivae, me cichorea, me malvae" ("As for me, olives, endives, and mallows provide sustenance").[6] Lord Monboddo describes his translation of an ancient epigram that demonstrates malva was planted upon the graves of the ancients, stemming from the belief that the dead could feed on such perfect plants.[7]

Mallow, which grows wild in the Middle East, is widely used as a source of nourishment in wartime and periods of austerity. Known as خبّيزة (*khubeza*) in Arabic, it is used as the main ingredient in a traditional Arab dish called by the same name; as well as in salads, soups and other dishes by the local Arab people.

## Species list

- *Malva aegyptia*
- *Malva aethiopica* C.J.S. Davis[8]
- *Malva alcea* L.—Greater Musk-mallow, Vervain Mallow
- *Malva assurgentiflora*
- *Malva brasiliensis* Desr.—Brazilian Mallow
- *Malva canariensis*
- *Malva cathayensis*
- *Malva cretica*
- *Malva dendromorpha*—Tree Mallow
- *Malva hispanica*
- *Malva microcarpa*
- *Malva microphylla*
- *Malva pacifica*
- *Malva parviflora* L.—Least Mallow, Cheeseweed, Cheeseweed Mallow, Small-whorl Mallow
- *Malva preissiana*—Australian Hollyhock
- *Malva pseudolavatera*
- *Malva pusilla*—Small Mallow
- *Malva qaiseri*
- *Malva rotundifolia* L.—Low Mallow
- *Malva stipulacea*
- *Malva subovata*
- *Malva sylvestris* L.—Common Mallow, High Mallow
- *Malva transcaucasica*
- *Malva tournefortiana*

- *Malva mohileviensis*
- *Malva moschata* L.—Musk-mallow
- *Malva neglecta*—Dwarf Mallow, Buttonweed, Cheeseplant, Cheeseweed, Common Mallow, Roundleaf Mallow
- *Malva nicaeensis* All.—French Mallow, Bull Mallow

- *Malva trifida*
- *Malva verticillata* L.—Chinese Mallow, Cluster Mallow

Sources:[3][9][10][11][12][13][14]

# References

[1]  "*Malva* L." (http://botany.si.edu/ing/INGsearch.cfm?searchword=Malva). *Index Nominum Genericorum*. International Association for Plant Taxonomy. 1996-02-09. . Retrieved 2008-05-09.

[2]  "*Malva* L." (http://www.ars-grin.gov/cgi-bin/npgs/html/genus.pl?7216). *Germplasm Resources Information Network*. United States Department of Agriculture. 2007-03-12. . Retrieved 2010-02-16.

[3]  Malvaceae Info: *Malva* (http://www.malvaceae.info/Genera/Malva/Malva.html)

[4]  http://www.etymonline.com/index.php?term=mallow

[5]  http://en.wiktionary.org/wiki/malva

[6]  Horace, *Odes 31, ver 15*, c. 30 BC

[7]  Letter from Monboddo to John Hope, 29 April 1779; reprinted by William Knight 1900 ISBN 1-85506-207-0.

[8]  C.J.S. Davis, *Malva aethiopica*, a new name for *Lavatera abyssinica* (Malvaceae): an endemic species of the Ethiopian Highlands, Phytotaxa 13: 56–58 (2010)

[9]  Africal Flowering Plants Database: *Malva* (http://www.ville-ge.ch/cjb/bd/africa/resultat.php)

[10]  Flora Europaea: *Malva* (http://rbg-web2.rbge.org.uk/cgi-bin/nph-readbtree.pl/feout?FAMILY_XREF=&GENUS_XREF=Malva& SPECIES_XREF=&TAXON_NAME_XREF=&RANK=species)

[11]  Flora of Pakistan: *Malva* (http://www.efloras.org/florataxon.aspx?flora_id=5&taxon_id=119571)

[12]  Flora of China: *Malva* checklist (http://www.efloras.org/florataxon.aspx?flora_id=3&taxon_id=119571)

[13]  "*Malva* L." (http://www.itis.gov/servlet/SingleRpt/SingleRpt?search_topic=TSN&search_value=21832). Integrated Taxonomic Information System. . Retrieved 9 May 2008.

[14]  UniProt. "Genus **Malva**" (http://beta.uniprot.org/taxonomy/96479). . Retrieved 2008-05-09.

# Malvaceae

<table>
<tr><td colspan="2" align="center">Malvaceae</td></tr>
<tr><td colspan="2" align="center">Least Mallow, Malva parviflora</td></tr>
<tr><td colspan="2" align="center">Scientific classification</td></tr>
<tr><td>Kingdom:</td><td>Plantae</td></tr>
<tr><td>(unranked):</td><td>Angiosperms</td></tr>
<tr><td>(unranked):</td><td>Eudicots</td></tr>
<tr><td>(unranked):</td><td>Rosids</td></tr>
<tr><td>Order:</td><td>Malvales</td></tr>
<tr><td>Family:</td><td>Malvaceae<br>Juss.</td></tr>
<tr><td colspan="2" align="center">Subfamilies</td></tr>
<tr><td colspan="2">Bombacoideae<br>Brownlowioideae<br>Byttnerioideae<br>Dombeyoideae<br>Grewioideae<br>Helicteroideae<br>Malvoideae<br>Sterculioideae<br>Tilioideae</td></tr>
</table>

**Malvaceae**, or the **mallow family**, is a family of flowering plants containing over 200 genera with close to 2,300 species.[1] Well known members of this family include okra, jute and cacao. The largest genera in terms of number of species include *Hibiscus* (300 species), *Sterculia* (250 species), *Dombeya* (225 species), *Pavonia* (200 species) and *Sida* (200 species).

## Taxonomy and nomenclature

The circumscription of the Malvaceae is very controversial. The traditional Malvaceae *sensu stricto* comprises a very homogeneous and cladistically monophyletic group. Another major circumscription, Malvaceae *sensu lato*, has been more recently defined on the basis that molecular techniques have shown that the commonly recognised families Bombacaceae, Tiliaceae, and Sterculiaceae, which have always been considered closely allied to Malvaceae *s.s.*, are not monophyletic groups. Thus the Malvaceae can be been expanded to include all of these families so as to compose a monophyletic group. Adopting this circumscription, Malvaceae incorporates a much larger number of genera.

This article is based on the second circumscription, as presented by the Angiosperm Phylogeny Website.[2] The Malvaceae *s.l.* (hereafter simply "Malvaceae") comprise nine subfamilies. A tentative cladogram of the family is shown below. The diamond denotes a poorly supported branching (<80%).

**Byttnerioideae**: 26 genera, 650 species. Pantropical, especially South America

**Grewioideae**: 25 genera, 770 species. Pantropical.

**Sterculioideae**: 12 genera, 430 species. Pantropical

**Tilioideae**: 3 genera, 50 species. Northern temperate regions and Central America

**Dombeyoideae**: About 20 genera, c.380 species. Palaeotropical, especially Madagascar and Mascarenes

♦ **Brownlowioideae**: 8 genera, c.70 species. Especially palaeotropical.

**Helicteroideae**: 8 to 12 genera, 10 to 90 species. Tropical, especially south east Asia.

♦ **Malvoideae**: 78 genera, 1,670 species. Temperate to tropical.

**Bombacoideae**: 12 genera, 120 species. Tropical, especially Africa and America

It is important to point out the relationships between these subfamilies are still either poorly supported or almost completely obscure, so that the circumscription of the family may change dramatically as new studies are published.

If looking for information about the traditional Malvaceae *s.s.*, we recommend referring to *Malvoideae*, the subfamily that approximately corresponds to that group.

The English common name 'mallow' (also applied to other members of Malvaceae) comes from Latin *malva* (also the source for the English word "mauve"). *Malva* itself was ultimately derived from the word for the plant in ancient Mediterranean languages.[3] Cognates of the word include Ancient Greek μαλάχη (*malákhē*) or μολόχη (*molókhē*), Modern Greek μολόχα (*molóha*), modern Arabic: ملوخية (*mulukhiyah*) and modern Hebrew: מלוחייה (*molokhia*).[3][4]

# Description

*Pavonia odorata*

Most species are herbs or shrubs but some are trees and lianas.

## Leaves and stems

Leaves are generally alternate, often palmately lobed or compound and palmately veined. The margin may be entire, but when dentate a vein ends at the tip of each tooth (*malvoid teeth*). Stipules are present. The stems contain mucous canals and often also mucous cavities. Hairs are common, and are most typically stellate.

## Flowers

The flowers are commonly borne in definite or indefinite axillary inflorescences, which are often reduced to a single flower, but may also be cauliflorous, oppositifolious or terminal. They often bear supernumerary bracts. They can be unisexual or bisexual and are generally actinomorphic, often associated with conspicuous bracts, forming an epicalyx. They generally have five valvate sepals, most frequently basally connate. Five imbricate petals. The stamens are five to numerous, connate at least at their bases, but often forming a tube around the pistils. The pistils are composed of two to many connate carpels. The ovary is superior, with axial placentation. Capitate or lobed stigma. The flowers have nectaries made of many tightly packed glandular hairs, usually positioned on the sepals.

Stellate hairs on the underside of a dried leaf of
*Malva alcea*

## Fruits

Most often a loculicidal capsule, a schizocarp or nut.

## Pollination

Self pollination is often avoided by means of protandry. Most species are entomophilous (pollinated by insects).

## Importance

A number of species are pests in agriculture, including *Abutilon theophrasti* and *Modiola caroliniana*, and others that are garden escapes. Cotton (4 species of *Gossypium*), kenaf (*Hibiscus cannabinus*), cacao, kola nut and okra (*Abelmoschus esculentus*) are important agricultural crops. The fruit and leaves of baobabs are edible, as is the fruit of the durian.

Durian fruits.

# References

[1] Judd & al.

[2] Angiosperm Phylogeny Website

[3] Douglas Harper. "mallow" (http://www.etymonline.com/index.php?term=mallow&allowed_in_frame=0). Online Etymology Dictionary. . Retrieved February 3, 2012.

[4] Khalid. "Molokheya: an Egyptian National Dish" (http://baheyeldin.com/egypt/molokheya-an-egyptian-national-dish.html). THe Baheyeldin Dynasty. . Retrieved September 10, 2011.

- Baum, D. A., W. S. Alverson, and R. Nyffeler (1998). "A durian by any other name: taxonomy and nomenclature of the core Malvales". *Harvard Papers in Botany* **3**: 315–330.
- Baum, D. A.; Dewitt Smith, S.; Yen, A.; Alverson, W. S.; Nyffeler, R.; Whitlock, B. A.; Oldham, R. L. (2004). "Phylogenetic relationships of Malvatheca (Bombacoideae and Malvoideae; Malvaceae sensu lato) as inferred from plastid DNA sequences". *American Journal of Botany* **91** (11): 1863–71. doi:10.3732/ajb.91.11.1863. PMID 21652333.
- Bayer, C. (1999). "Support for an expanded family concept of Malvaceae within a recircumscribed order Malvales: a combined analysis of plastidatpB andrbcL DNA sequences". *Botanical Journal of the Linnean Society* **129** (4): 267–303. doi:10.1006/bojl.1998.0226.
- Bayer, C. and K. Kubitzki 2003. Malvaceae, pp. 225-311. In K. Kubitzki (ed.), *The Families and Genera of Vascular Plants*, vol. 5, Malvales, Capparales and non-betalain Caryophyllales.
- Edlin, H. L. (1935). "A Critical Revision of Certain Taxonomic Groups of the Malvales Part Ii1". *New Phytologist* **34** (2): 122–143. doi:10.1111/j.1469-8137.1935.tb06834.x.
- Judd, W. S.; Manchester, S. R. (1997). "Circumscription of Malvaceae (Malvales) as Determined by a Preliminary Cladistic Analysis of Morphological, Anatomical, Palynological, and Chemical Characters". *Brittonia* **49** (3): 384–405. doi:10.2307/2807839. JSTOR 2807839.
- Judd, W. S., C. S. Campbell, E. A. Kellogg and P. F. Stevens. Plant Systematics: A Phylogenetic Approach.
- Maas, P. J. M. and L. Y. Th. Westra. 2005. *Neotropical Plant Families* (3rd edition).
- Perveen, A.; Grafstrom, E.; El-Ghazaly, G. (2004). "World Pollen and Spore Flora 23. Malvaceae Adams. P.p. Subfamilies: Grewioideae, Tilioideae, Brownlowioideae". *Grana* **43** (3): 129. doi:10.1080/00173130410000730. ISBN 3130410000730.
- Tate, J. A., J. F. Aguilar, S. J. Wagstaff, J. C. La Duke, T. A. Bodo Slotta and B. B. Simpson (2005). "Phylogenetic relationships within the tribe Malveae (Malvaceae, subfamily Malvoideae) as inferred from ITS sequence data". *American Journal of Botany* **92** (4): 584–602. doi:10.3732/ajb.92.4.584. PMID 21652437. (abstract online here (http://www.amjbot.org/cgi/content/abstract/92/4/584)).

# External links

- Malvaceae in Topwalks (http://www.topwalks.net/plants/generos/malvaceae_02.htm)
- Alverson, William S., Barbara A. Whitlock, Reto Nyffeler, Clemens Bayer and David A. Baum. 1999. Phylogeny of the core Malvales: evidence from *ndh*F sequence data. *American Journal of Botany* 86: 1474-1486. (http://www.amjbot.org/cgi/content/abstract/86/10/1474)
- Core Malvales (http://tolweb.org/tree?group=Core_Malvales&contgroup=Malvales) from Tree of Life (http://tolweb.org)
- Malvaceae: plants of Hawaii (image gallery) (http://www.hear.org/starr/hiplants/images/family/malvaceae.htm) from HEAR (http://www.hear.org/)
- Malvaceae Gallery (http://www.malvaceae.info/Genera/gallery.html)
- Malvaceae of Mongolia in FloraGREIF (http://greif.uni-greifswald.de/floragreif/?fam=Malvaceae&gen=&spec=&flora_search=taxon)

# Byttnerioideae

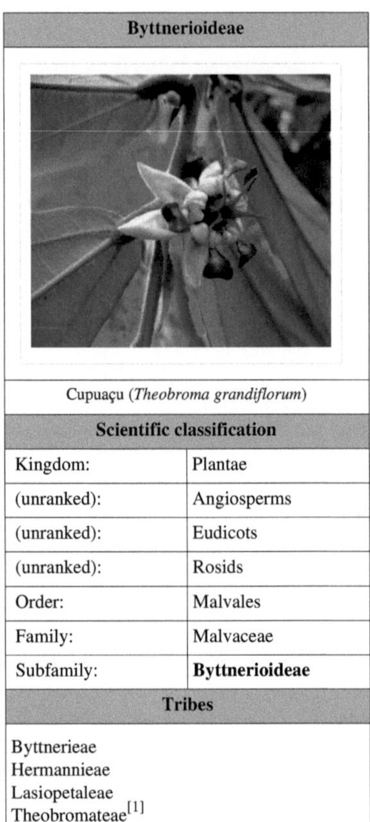

Cupuaçu (*Theobroma grandiflorum*)

| Scientific classification | |
| --- | --- |
| Kingdom: | Plantae |
| (unranked): | Angiosperms |
| (unranked): | Eudicots |
| (unranked): | Rosids |
| Order: | Malvales |
| Family: | Malvaceae |
| Subfamily: | **Byttnerioideae** |

| Tribes |
| --- |
| Byttnerieae<br>Hermannieae<br>Lasiopetaleae<br>Theobromateae[1] |

**Byttnerioideae** is a subfamily of the flowering plant family Malvaceae.[1]

## Genera

Tribe Byttnerieae

- *Abroma* Jacq.
- *Ayenia* L.
- *Byttneria* Loefl.
- *Kleinhovia* L.
- *Leptonychia* Turcz.
- *Megatritheca* Cristóbal
- *Rayleya* Cristóbal
- *Scaphopetalum* Mast.[2]

Tribe Hermannieae

- *Dicarpidium* F.Muell.
- *Gilesia* F.Muell.
- *Hermannia* L.
- *Melochia* L.
- *Waltheria* L.[3]

Tribe Lasiopetaleae

- *Commersonia* J.R.Forst. & G.Forst.
- *Guichenotia* J.Gay
- *Hannafordia* F.Muell.
- *Keraudrenia* J.Gay
- *Lasiopetalum* Sm.
- *Lysiosepalum* F.Muell.
- *Rulingia* R.Br.
- *Seringia* J.Gay

- *Thomasia* J.Gay[4]

Tribe Theobromateae

- *Glossostemon* Desf.
- *Guazuma* Mill.
- *Herrania* Goudot
- *Theobroma* L.[5]

Incertae sedis

- *Maxwellia* Baill.[6]

# References


[1] "Family: *Malvaceae* Juss., nom. cons. subfam. *Byttnerioideae*" (http://www.ars-grin.gov/cgi-bin/npgs/html/family.pl?1750). *Germplasm Resources Information Network*. United States Department of Agriculture. 2007-04-12. . Retrieved 2011-02-19.

[2] "GRIN Genera of *Malvaceae* tribe *Byttnerieae*" (http://www.ars-grin.gov/cgi-bin/npgs/html/gnlist.pl?1759). *Germplasm Resources Information Network*. United States Department of Agriculture. . Retrieved 2011-02-19.

[3] "GRIN Genera of *Malvaceae* tribe *Hermannieae*" (http://www.ars-grin.gov/cgi-bin/npgs/html/gnlist.pl?1762). *Germplasm Resources Information Network*. United States Department of Agriculture. . Retrieved 2011-02-19.

[4] "GRIN Genera of *Malvaceae* tribe *Lasiopetaleae*" (http://www.ars-grin.gov/cgi-bin/npgs/html/gnlist.pl?1761). *Germplasm Resources Information Network*. United States Department of Agriculture. . Retrieved 2011-02-19.

[5] "GRIN Genera of *Malvaceae* tribe *Theobromateae*" (http://www.ars-grin.gov/cgi-bin/npgs/html/gnlist.pl?1761). *Germplasm Resources Information Network*. United States Department of Agriculture. . Retrieved 2011-02-19.

[6] "GRIN Genera of *Malvaceae* subfam. *Byttnerioideae*" (http://www.ars-grin.gov/cgi-bin/npgs/html/gnlist.pl?1750). *Germplasm Resources Information Network*. United States Department of Agriculture. . Retrieved 2011-02-19.

# Malvales

| Malvales | |
|---|---|
| 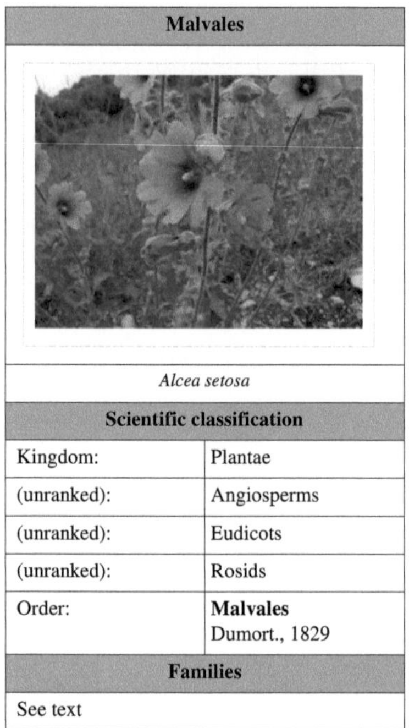<br>*Alcea setosa* | |
| **Scientific classification** | |
| Kingdom: | Plantae |
| (unranked): | Angiosperms |
| (unranked): | Eudicots |
| (unranked): | Rosids |
| Order: | **Malvales**<br>Dumort., 1829 |
| **Families** | |
| See text | |

**Malvales** are an order of flowering plants. As circumscribed by APG II-system, it includes about 6000 species within nine families. The order is placed in the **eurosids II**, which are part of the eudicots.

The plants are mostly shrubs and trees; most of its families have a cosmopolitan distribution in the tropics and subtropics with limited expansion into temperate regions. An interesting distribution occurs in Madagascar, where there are three endemic families of Malvales (Sphaerosepalaceae, Sarcolaenaceae and Diegodendraceae).

*Hibiscus moscheutos*

Many species of Malvaceae s.l. are known for their wood, with that of *Ochroma* (balsa) being known for its lightness, and that of *Tilia* (lime, linden, or basswood) as a popular wood for carving. The cacao tree (*Theobroma cacao*) is used as an ingredient for chocolate. Kola nuts (genus *Cola*) are notable for their high content of caffeine, and in past were commonly used for preparing of various cola drinks. Other well-known members of Malvales in the APG II sense are daphnes, hibiscus, hollyhocks, okra, baobab trees, cotton, and kapok.

The morphology of Malvales is diverse, and there are few common characteristics. Among those most commonly encountered are palmate leaves, connate sepals, and a specific structure and chemical composition of the seeds. The cortex is often fibrous, built of soft phloem layers.

Families (APG system)

- Bixaceae

- Cistaceae
- Cochlospermaceae
- Cytinaceae[1]
- Diegodendraceae
- Dipterocarpaceae
- Malvaceae
- Muntingiaceae
- Neuradaceae
- Sarcolaenaceae
- Sphaerosepalaceae
- Thymelaeaceae

## Classification

Family boundaries and circumscriptions of the "core Malvales" families Malvaceae, Bombacaceae, Tiliaceae, and Sterculiaceae have long been problematic. A close relationship among these families, and particularly Malvaceae and Bombacaceae, has generally been recognized although until recently most classification systems have maintained them as separate families. With numerous molecular phylogenies showing that Sterculiaceae, Bombacaceae, and Tiliaceae as traditionally defined are either paraphyletic or polyphyletic, a consensus has been emerging that there has been a trend to expand Malvaceae to include these three families. This expanded circumscription of Malvaceae has been recognized in the most recent version of the Thorne system, by the Angiosperm Phylogeny Group, and in the most recent comprehensive treatment of vascular plant families and genera, the Kubitzki system (Bayer and Kubitzki, 2003).

The dominant family in the APG II-system is the extended Malvaceae (Malvaceae *sensu lato*) with over 4000 species, followed by Thymelaeaceae with 750 species. This expanded circumscription of Malvaceae is taken to include the families Bombacaceae, Sterculiaceae and Tiliaceae. Under the older Cronquist system the order contained these four "core Malvales" families plus the Elaeocarpaceae and was placed among the Dilleniidae. Some of the currently included families were placed by Cronquist in the Violales.

## References

[1] Nickrent, Daniel L. "Cytinaceae are sister to Muntingiaceae (Malvales)", *Taxon* 56 (4): 1129-1135 (2007) ( abstract (http://www. ingentaconnect.com/content/iapt/tax/2007/00000056/00000004/art00011))

- Alverson, W. S., K. G. Karol, D. A. Baum, M. W. Chase, S. M. Swensen, R. McCourt, and K. J. Sytsma (1998). Circumscription of the Malvales and relationships to other Rosidae: Evidence from rbcL sequence data. *American Journal of Botany* **85**, 876-887. (Available online: Abstract (http://www.amjbot.org/cgi/content/abstract/85/6/876))
- Bayer, C. and K. Kubitzki. 2003. Malvaceae, pp. 225–311. In K. Kubitzki (ed.), *The Families and Genera of Vascular Plants*, vol. 5, Malvales, Capparales and non-betalain Caryophyllales.
- Edlin, H. L. 1935. A critical revision of certain taxonomic groups of the Malvales. *New Phytologist* 34: 1-20, 122-143.
- Judd, W.S., C. S. Campbell, E. A. Kellogg, P. F. Stevens, M. J. Donoghue (2002). *Plant Systematics: A Phylogenetic Approach, 2nd edition*. pp. 405–410 (Malvales). Sinauer Associates, Sunderland, Massachusetts. ISBN 0-87893-403-0.
- Kubitzki, K. and M. W. Chase. 2003. Introduction to Malvales, pp. 12– 16. In K. Kubitzki (ed.), *The Families and Genera of Vascular Plants*, vol. 5, Malvales, Capparales and non-betalain Caryophyllales.
- du Mortier, B. C. J. (1829). *Analyse des Familles de Plantes, avec l'indication des principaux genres qui s'y rattachent*, p. 43. Imprimerie de J. Casterman, Tournay.

- Watson, L., and Dallwitz, M. J. (1992 onwards). The families of flowering plants: descriptions, illustrations, identification, and information retrieval (http://delta-intkey.com/angio/). http://delta-intkey.com
- Whitlock, B. A. (October 2001). Malvales (Mallow). In: *Nature Encyclopedia of Life Sciences*. Nature Publishing Group, London. (Available online: DOI (http://dx.doi.org/10.1038/npg.els.0003727) | ELS site (http://www.els.net/))


## External links

- Tree of Life Malvales (http://tolweb.org/Malvales/21050)

# Brownlowioideae

| Brownlowioideae | |
|---|---|
| **Scientific classification** | |
| Kingdom: | Plantae |
| (unranked): | Angiosperms |
| (unranked): | Eudicots |
| (unranked): | Rosids |
| Order: | Malvales |
| Family: | Malvaceae |
| Subfamily: | Brownlowioideae |
| **Genera** | |
| • *Berrya* <br> • *Brownlowia* <br> • *Carpodiptera* <br> • *Christiana* <br> • *Diplodiscus* <br> • *Hainania* <br> • *Indagator* <br> • *Jarandersonia* <br> • *Pentace* <br> • *Pityranthe* | |

**Brownlowioideae** is a subfamily of the botanical family Malvaceae. The genera in this subfamily used to be a part of the paraphyletic Tiliaceae until taxonomic revisions in part by the APG II system.

# Rosids

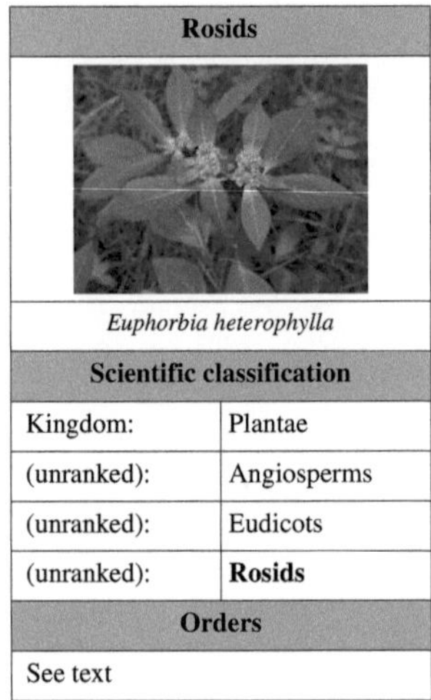

**Rosids**

*Euphorbia heterophylla*

| Scientific classification | |
|---|---|
| Kingdom: | Plantae |
| (unranked): | Angiosperms |
| (unranked): | Eudicots |
| (unranked): | **Rosids** |
| **Orders** | |
| See text | |

The **rosids** are members of a large clade of flowering plants, containing about 70,000 species,[1] more than a quarter of all angiosperms.[2] The clade is divided into 16 to 20 orders, depending upon circumscription and classification. These orders, in turn, together comprise about 140 families.[3] The rosids and the asterids are by far the largest clades in the eudicots.

Fossil rosids are known from the Cretaceous period. Molecular clock estimates indicate that the rosids originated in the Aptian or Albian stages of the Cretaceous, between 125 and 99.6 million years ago.[4][5]

## The name

The name "rosids" is based upon the name "Rosidae", which had usually been understood to be a subclass. In 1967, Armen Takhtajan showed that the correct basis for the name "Rosidae" is a description of a group of plants published in 1830 by Friedrich Gottlieb Bartling.[6] This clade was later renamed "Rosidae" and has been variously delimited by different authors. The name "rosids" is informal, and not assumed to have any particular taxonomic rank like the names authorized by the ICBN. The rosids are monophyletic based upon evidence found by molecular phylogenetic analysis.

Three different definitions of the rosids are currently in use. Some authors include the orders Saxifragales and Vitales in the rosids.[7] Others exclude both of these orders.[8] The circumscription used in this article is that of the APG II classification, which includes Vitales, but excludes Saxifragales.

## Relationships

The rosids and Saxifragales form a clade.[1][8] This is one of 6 groups that compose the Pentapetalae (core eudicots minus Gunnerales),[9] the others being Berberidopsidales, Caryophyllales, Dilleniales, Santalales, and the asterids. Almost nothing is known about the relationships between these groups.

## Classification

The rosids consist of two groups: the order Vitales and the eurosids (true rosids). The eurosids, in turn are divided into seven groups: Fabidae, Geraniales, Myrtales, Crossosomatales, Picramniales,[8] Malvidae,[9] and the unplaced family Apodanthaceae.[10] The Fabidae are often called the fabids, or eurosids I. Likewise, the Malvidae are often called the malvids, or eurosids II.

## Orders

The rosids consist of 17 orders and 2 families that are placed incertae sedis (not in any order). In addition to Vitales, Geraniales, Myrtales, Crossosomatales, and Picramniales, there are 8 orders in Fabidae and 4 orders in Malvidae. In 2009, Hengchang Wang and co-authors proposed that Malvidae be expanded to include Geraniales, Myrtales, Crossosomatales, and Picramniales. This larger circumscription of Malvidae received strong statistical support (100% bootstrap percentage) in their analysis. Some of the orders have only recently been recognized.[8] These are Vitales,[11] Zygophyllales,[12] Crossosomatales,[13] Picramniales,[14] and Huerteales.[15]

### Unplaced families

The families Apodanthaceae and Huaceae are included in the rosids, but not placed in any of its orders.

Apodanthaceae is an enigmatic family of achlorophyllous parasites. They have been provisionally placed in Cucurbitales by some,[8] but their affinities remain obscure.[10] The chloroplast genes that have been used to infer plant phylogeny do not provide much phylogenetic information for plants that lack chlorophyll, because in this case, these genes are nonfunctional pseudogenes.

The family Huaceae is a member of the COM (Celastrales, Oxalidales, Malpighiales) clade of Fabidae. The question about Huaceae is whether it should be included in one of the COM orders or in an order by itself as a 4th member of the COM clade. Two studies have indicated that it should be placed in Oxalidales,[16][17] while one has indicated that it should not.[1]

## Phylogeny

The phylogeny shown below is adapted from Wang and co-authors (2009),[1] with order names from the Angiosperm Phylogeny Website.[8] Branches with less than 50% bootstrap support are collapsed. Other branches have 100% bootstrap support except where shown.

Vitales

Zygophyllales

COM clade
Celastrales
Oxalidales
Malpighiales

Fabidae
Fabales

Rosales
nitrogen-fixing clade
Fagales
Cucurbitales

eurosids

65% Geraniales
Myrtales

Crossosomatales

Picramniales

Malvidae sensu lato
Sapindales

Huerteales
Malvidae sensu stricto
Brassicales
Malvales

The *nitrogen-fixing clade* contains a high number of actinorhizal plants (which have root nodules containing nitrogen fixing bacteria, helping the plant grow in poor soils). Not all plants in this clade are actinorhizal, however.

# References

[1]  Hengchang Wang, Michael J. Moore, Pamela S. Soltis, Charles D. Bell, Samuel F. Brockington, Roolse Alexandre, Charles C. Davis, Maribeth Latvis, Steven R. Manchester, and Douglas E. Soltis (10Mar2009), "Rosid radiation and the rapid rise of angiosperm-dominated forests", *Proceedings of the National Academy of Sciences* **106** (10): 3853–3858, doi:10.1073/pnas.0813376106, PMC 2644257, PMID 19223592

[2]  Robert W. Scotland and Alexandra H. Wortley (2003), "How many species of seed plants are there?", *Taxon* **52** (1): 101–104, doi:10.2307/3647306, JSTOR 3647306

[3]  Douglas E. Soltis, Pamela S. Soltis, Peter K. Endress, and Mark W. Chase (2005), *Phylogeny and Evolution of the Angiosperms*, Sunderland, MA, USA: Sinauer, ISBN 978-0-87893-817-9

[4]  Davies, T.J., Barraclough, T.G., Chase, M.W., Soltis, P.S., Soltis, D.E., and Savolainen, V. (2004), "Darwin's abominable mystery: Insights from a supertree of the angiosperms", *Proceedings of the National Academy of Sciences* **101** (7): 1904–1909, Bibcode 2004PNAS..101.1904D, doi:10.1073/pnas.0308127100, PMC 357025, PMID 14766971

[5]  Susana Magallón and Amanda Castillo (2009), "Angiosperm diversification through time", *American Journal of Botany* **96** (1): 349–365, doi:10.3732/ajb.0800060, PMID 21628193

[6]  James L. Reveal (2008 onward), "A Checklist of Family and Suprafamilial Names for Extant Vascular Plants" (http://www. plantsystematics.org/reveal/pbio/fam/supgennames.html), *Home page of James L. Reveal and C. Rose Broome* (http://www. plantsystematics.org/reveal),

[7]  J. Gordon Burleigh, Khidir W. Hilu, and Douglas E. Soltis (2009), *File 7* (http://www.biomedcentral.com/content/supplementary/ 1471-2148-9-61-S7.pdf), "Inferring phylogenies with incomplete data sets: a 5-gene, 567-taxon analysis of angiosperms", *BMC Evolutionary Biology* **9**: 61, doi:10.1186/1471-2148-9-61, PMC 2674047, PMID 19292928,

[8]  Peter F. Stevens (2001 onwards), *Angiosperm Phylogeny Website* (http://www.mobot.org/MOBOT/Research/APweb/welcome.html),

[9]  Philip D. Cantino, James A. Doyle, Sean W. Graham, Walter S. Judd, Richard G. Olmstead, Douglas E. Soltis, Pamela S. Soltis, and Michael J. Donoghue (2007), "Towards a phylogenetic nomenclature of *Tracheophyta*" (http://www.phylodiversity.net/donoghue/publications/ MJD_papers/2007/164_Cantino_Taxon07.pdf), *Taxon* **56** (3): 822–846, doi:10.2307/25065865,

[10]  Daniel L. Nickrent, "Apodanthaceae" (http://www.parasiticplants.siu.edu/Apodanthaceae/index.html), *The Parasitic Plant Connection* (http://www.parasiticplants.siu.edu/),

[11]  James L. Reveal. (1995). page 72 in Newly required suprageneric names in vascular plants. Phytologia 79(2):68-76

[12]  Chalk, L. 1983. Wood structure. Pp. 1-51 [1-2 by C. R. Melcalfe], in Metcalfe, C. R., & Chalk, L., Anatomy of the Dicotyledons, Second Edition. Volume II. Wood Structure and Conclusion of the General Introduction. Clarendon Press, Oxford. ISBN 978-0-19-854559-0.

[13]  Klaus Kubitzki (2007), "Introduction to Crossosomatales", in Klaus Kubitzki, *The Families and Genera of Vascular Plants, vol.IX*, Berlin,Heidelberg: Springer-Verlag

[14]  John Hutchinson *The Families of Flowering Plants* 3rd edition. 1973. Oxford University Press.

[15]  Andreas Worberg, Mac H. Alford, Dietmar Quandt, and Thomas Borsch (2009), "Huerteales sister to Brassicales plus Malvales, and newly circumscribed to include Dipentodon, Gerrardina, Huertea, Perrottetia, and Tapiscia", *Taxon* **58** (2): 468–478

[16]  Douglas E. Soltis, Matthew A. Gitzendanner, and Pamela S. Soltis (2007), "A 567-taxon data set for angiosperms: The challenges posed by Bayesian analyses of large data sets", *International Journal of Plant Sciences* **168** (2): 137–157, doi:10.1086/509788

[17]  Li-Bing Zhang and Mark P. Simmons (2006), "Phylogeny and delimitation of the Celastrales inferred from nuclear and plastid genes", *Systematic Botany* **31** (1): 122–137, doi:10.1600/036364406775971778

# Plant

<table>
<tr><td colspan="2" align="center">Plants<br>Temporal range:<br>Early Cambrian to recent, but see text,</td></tr>
<tr><td colspan="2" align="center"></td></tr>
<tr><td colspan="2" align="center">Scientific classification</td></tr>
<tr><td>Domain:</td><td>Eukaryota</td></tr>
<tr><td>(unranked):</td><td>Archaeplastida</td></tr>
<tr><td>Kingdom:</td><td>Plantae<br>Haeckel, 1866[1]</td></tr>
<tr><td colspan="2" align="center">Divisions</td></tr>
<tr><td colspan="2">

**Green algae**

- Chlorophyta
- Charophyta

**Land plants (embryophytes)**

- **Non-vascular land plants (bryophytes)**
  - Marchantiophyta—liverworts
  - Anthocerotophyta—hornworts
  - Bryophyta—mosses
  - †Horneophytopsida
- **Vascular plants (tracheophytes)**
  - †Rhyniophyta—rhyniophytes
  - †Zosterophyllophyta—zosterophylls
  - Lycopodiophyta—clubmosses
  - †Trimerophytophyta—trimerophytes
  - Pteridophyta—ferns and horsetails
  - †Progymnospermophyta
  - **Seed plants (spermatophytes)**
    - †Pteridospermatophyta—seed ferns
    - Pinophyta—conifers
    - Cycadophyta—cycads
    - Ginkgophyta—ginkgo
    - Gnetophyta—gnetae
    - Magnoliophyta—flowering plants

**†Nematophytes**

</td></tr>
</table>

**Plants** are living organisms belonging to the kingdom **Plantae**. Precise definitions of the kingdom vary, but as the term is used here, plants include familiar organisms such as flowering plants, conifers, ferns, mosses, and green algae, but do not include seaweeds like kelp, nor fungi and bacteria. The group is also called **green plants** or **Viridiplantae** in Latin. They obtain most of their energy from sunlight via photosynthesis using chlorophyll contained in chloroplasts, which gives them their green color. Some plants are parasitic and may not produce normal amounts of chlorophyll or photosynthesize.

Precise numbers are difficult to determine, but as of 2010, there are thought to be 300–315 thousand species of plants, of which the great majority, some 260–290 thousand, are seed plants (see the table below).[2]

The scientific study of plants is known as botany.

# Definition

Plants are one of the two groups into which all living things have been traditionally divided; the other is animals. The division goes back at least as far as Aristotle (384 BC – 322 BC) who distinguished between plants which generally do not move, and animals which often are mobile to catch their food. Much later, when Linnaeus (1707–1778) created the basis of the modern system of scientific classification, these two groups became the kingdoms Vegetabilia (later Metaphyta or Plantae) and Animalia (also called Metazoa). Since then, it has become clear that the plant kingdom as originally defined included several unrelated groups, and the fungi and several groups of algae were removed to new kingdoms. However, these organisms are still often considered plants, particularly in popular contexts.

Outside of formal scientific contexts, the term "plant" implies an association with certain traits, such as being multicellular, possessing cellulose, and having the ability to carry out photosynthesis.[3][4]

## Current definitions of Plantae

When the name Plantae or plant is applied to a specific group of organisms or taxon, it usually refers to one of three concepts. From least to most inclusive, these three groupings are:

| Name(s) | Scope | Description |
| --- | --- | --- |
| Land plants, also known as Embryophyta or Metaphyta. | Plantae *sensu strictissimo* | This group includes the liverworts, hornworts, mosses, and vascular plants, as well as fossil plants similar to these surviving groups. |
| **Green plants** - also known as **Viridiplantae**, **Viridiphyta** or **Chlorobionta** | Plantae *sensu stricto* | This group includes the land plants plus various groups of green algae, including stoneworts. The names given to these groups vary considerably as of July 2011. Viridiplantae encompass a group of organisms that possess chlorophyll *a* and *b*, have plastids that are bound by only two membranes, are capable of storing starch, and have cellulose in their cell walls. It is this clade which is mainly the subject of this article. |
| Archaeplastida, Plastida or Primoplantae | Plantae *sensu lato* | This group comprises the green plants above plus Rhodophyta (red algae) and Glaucophyta (glaucophyte algae). This clade includes the organisms that eons ago acquired their chloroplasts directly by engulfing cyanobacteria. |

Another way of looking at the relationships between the different groups which have been called "plants" is through a cladogram, which shows their evolutionary relationships. The evolutionary history of plants is not yet completely settled, but one accepted relationship between the three groups described above is shown below.[5] Those which have been called "plants" are in bold.

Glaucophyta (glaucophyte algae)

Rhodophyta (red algae)

Chlorophyta (part of green algae)

streptophyte algae (part of green algae)

**Archaeplastida**

**Viridiplantae**      Streptophyta      Charales (stoneworts, often included in green algae)

**land plants** or embryophytes

The way in which the groups of green algae are combined and named varies considerably between authors.

Many of the classification controversies involve organisms that are rarely encountered and are of minimal apparent economic significance, but are crucial in developing an understanding of the evolution of modern flora.

## Algae

Algae comprise several different groups of organisms which produce energy through photosynthesis and for that reason have been included in the plant kingdom in the past. Most conspicuous among the algae are the seaweeds, multicellular algae that may roughly resemble land plants, but are classified among the brown, red and green algae. Each of these algal groups also includes various microscopic and single-celled organisms. There is good evidence that some of these algal groups arose independently from separate non-photosynthetic ancestors, with the result that many groups of algae are no longer classified within the plant kingdom as it is defined here.[6][7]

The Viridiplantae, the green plants – green algae and land plants – form a clade, a group consisting of all the descendants of a common ancestor. With a few exceptions among the green algae, all green plants have many features in common, including cell walls containing cellulose, chloroplasts containing chlorophylls $a$ and $b$, and food stores in the form of starch. They undergo closed mitosis without centrioles, and typically have mitochondria with flat cristae. The chloroplasts of green plants are surrounded by two membranes, suggesting they originated directly from endosymbiotic cyanobacteria.

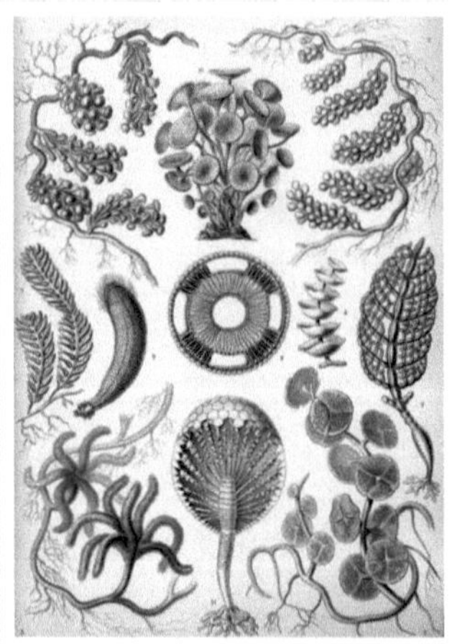

Green algae from Ernst Haeckel's *Kunstformen der Natur*, 1904.

Two additional groups, the Rhodophyta (red algae) and Glaucophyta (glaucophyte algae), also have chloroplasts which appear to be derived directly from endosymbiotic cyanobacteria, although they differ in the pigments which are used in photosynthesis and so are different in colour. All three groups together are generally believed to have a single common origin, and so are classified together in the taxon Archaeplastida, whose name implies that the chloroplasts or plastids of all the members of the taxon were derived from a single ancient endosymbiotic event. This is the broadest modern definition of the plants.

In contrast, most other algae (e.g. heterokonts, haptophytes, dinoflagellates, and euglenids) not only have different pigments but also have chloroplasts with three or four surrounding membranes. They are not close relatives of the Archaeplastida, presumably having acquired chloroplasts separately from ingested or symbiotic green and red algae. They are thus not included in even the broadest modern definition of the plant kingdom, although they were in the past.

The green plants or Viridiplantae were traditionally divided into the green algae (including the stoneworts) and the land plants. However, it is now known that the land plants evolved from within a group of green algae, so that the green algae by themselves are a paraphyletic group, i.e. a group which excludes some of the descendants of a common ancestor. Paraphyletic groups are generally avoided in modern classifications, so that in recent treatments the Viridiplantae have been divided into two clades, the Chlorophyta and the Streptophyta (or Charophyta).[8][9]

The Chlorophyta (a name that has also been used for *all* green algae) are the sister group to the group from which the land plants evolved. There are about 4,300 species[10] of mainly marine organisms, both unicellular and multicellular. The latter include the sea lettuce, *Ulva*.

The other group within the Viridiplantae are the mainly freshwater or terrestrial Streptophyta (or Charophyta), which consist of several groups of green algae plus the stoneworts and land plants. (The names have been used differently, e.g. Streptophyta to mean the group which excludes the land plants and Charophyta for the stoneworts alone or the stoneworts plus the land plants.) Streptophyte algae are either unicellular or form multicellular filaments, branched or unbranched.[9] The genus *Spirogyra* is a filamentous streptophyte alga familiar to many, as it is often used in teaching and is one of the organisms responsible for the algal "scum" which pond-owners so dislike. The freshwater stoneworts strongly resemble land plants and are believed to be their closest relatives. Growing underwater, they consist of a central stalk with whorls of branchlets, giving them a superficial resemblance to horsetails, species of the genus *Equisetum*, which are true land plants.

## Fungi

The classification of fungi has been controversial until quite recently in the history of biology. Linnaeus' original classification placed the fungi within the Plantae, since they were unquestionably not animals or minerals and these were the only other alternatives. With later developments in microbiology, in the 19th century Ernst Haeckel felt that another kingdom was required to classify newly discovered micro-organisms. The introduction of the new kingdom Protista in addition to Plantae and Animalia, led to uncertainty as to whether fungi truly were best placed in the Plantae or whether they ought to be reclassified as protists. Haeckel himself found it difficult to decide and it was not until 1969 that a solution was found whereby Robert Whittaker proposed the creation of the kingdom Fungi. Molecular evidence has since shown that the last common ancestor (concestor) of the Fungi was probably more similar to that of the Animalia than of any other kingdom, including the Plantae.

Whittaker's original reclassification was based on the fundamental difference in nutrition between the Fungi and the Plantae. Unlike plants, which generally gain carbon through photosynthesis, and so are called autotrophic phototrophs, fungi generally obtain carbon by breaking down and absorbing surrounding materials, and so are called heterotrophic saprotrophs. In addition, the substructure of multicellular fungi is different from that of plants, taking the form of many chitinous microscopic strands called hyphae, which may be further subdivided into cells or may form a syncytium containing many eukaryotic nuclei. Fruiting bodies, of which mushrooms are most familiar example, are the reproductive structures of fungi, and are unlike any structures produced by plants.

## Diversity

The table below shows some species count estimates of different green plant (Viridiplantae) divisions. It suggests there are about 300,000 species of living Viridiplantae, of which 85-90% are flowering plants. (Note: as these are from different sources and different dates, they are not necessarily comparable, and like all species counts, are subject to a degree of uncertainty in some cases.)

### Diversity of living green plant (Viridiplantae) divisions

| Informal group | Division name | Common name | No. of living species | Approximate No. in informal group |
|---|---|---|---|---|
| Green algae | **Chlorophyta** | green algae (chlorophytes) | 3,800 [11] – 4,300 [12] | 8,500 (6,600 - 10,300) |
| | **Charophyta** | green algae (e.g. desmids & stoneworts) | 2,800; [13] 4,000-6,000 [14] | |
| Bryophytes | **Marchantiophyta** | liverworts | 6,000-8,000 [15] | 19,000 (18,100 - 20,200) |
| | **Anthocerotophyta** | hornworts | 100-200 [16] | |
| | **Bryophyta** | mosses | 12,000 [17] | |
| Pteridophytes | **Lycopodiophyta** | club mosses | 1,200 [7] | 12,000 (12,200) |
| | **Pteridophyta** | ferns, whisk ferns & horsetails | 11,000 [7] | |
| Seed plants | **Cycadophyta** | cycads | 160 [18] | 260,000 (259,511) |
| | **Ginkgophyta** | ginkgo | 1 [19] | |
| | **Pinophyta** | conifers | 630 [7] | |
| | **Gnetophyta** | gnetophytes | 70 [7] | |
| | **Magnoliophyta** | flowering plants | 258,650 [20] | |

The naming of plants is governed by the International Code of Botanical Nomenclature and International Code of Nomenclature for Cultivated Plants (see cultivated plant taxonomy).

## Evolution

Further information: Evolutionary history of plants

The evolution of plants has resulted in increasing levels of complexity, from the earliest algal mats, through bryophytes, lycopods, ferns to the complex gymnosperms and angiosperms of today. The groups which appeared earlier continue to thrive, especially in the environments in which they evolved.

Evidence suggests that an algal scum formed on the land 1200 [21] million years ago, but it was not until the Ordovician Period, around 450.0 [22] million years ago, that land plants appeared.[23] However, new evidence from the study of carbon isotope ratios in Precambrian rocks has suggested that complex photosynthetic plants developed on the earth over 1000 m.y.a.[24] These began to diversify in the late Silurian Period, around 420 [25] million years ago, and the fruits of their diversification are displayed in remarkable detail in an early Devonian fossil assemblage from the Rhynie chert. This chert preserved early plants in cellular detail, petrified in volcanic springs. By the middle of the Devonian Period most of the features recognised in plants today are present, including roots, leaves and secondary wood, and by late Devonian times seeds had evolved.[26] Late Devonian plants had thereby reached a degree of sophistication that allowed them to form forests of tall trees. Evolutionary innovation continued after the Devonian period. Most plant groups were relatively unscathed by the Permo-Triassic extinction event, although the structures of communities changed. This may have set the scene for the evolution of flowering plants in the Triassic (~200 [27] million years ago), which exploded in the Cretaceous and Tertiary. The latest major group of plants to evolve were the grasses, which became important in the mid Tertiary, from around 40 [28] million years ago. The grasses, as well as many other groups, evolved new mechanisms of metabolism to survive the low $CO_2$ and warm, dry conditions of the tropics over the last 10 [29] million years.

A proposed phylogenetic tree of Plantae, after Kenrick and Crane,[30] is as follows, with modification to the Pteridophyta from Smith et al.[31] The Prasinophyceae may be a paraphyletic basal group to all green plants.

ophyceae (micromonads)

Streptobionta

Embryophytes

Stomatophytes

Polysporangiates

Tracheophytes

Eutracheophytes

Euphyllophytina

Lignophytia

**Spermatophytes** (seed plants)

Progymnospermophyta †

**Pteridophyta**

Pteridopsida (true ferns)

Marattiopsida

Equisetopsida (horsetails)

Psilotopsida (whisk ferns & adders'-tongues)

Cladoxylopsida †

Lycophytina

**Lycopodiophyta**

Zosterophyllophyta †

Rhyniophyta †

*Aglaophyton* †

Horneophytopsida †

**Bryophyta** (mosses)

**Anthocerotophyta** (hornworts)

**Marchantiophyta** (liverworts)

**Charophyta**

Trebouxiophyceae
(Pleurastrophyceae)

**rophyta**

Chlorophyceae

Ulvophyceae

## Embryophytes

The plants that are likely most familiar to us are the multicellular land plants, called embryophytes. They include the vascular plants, plants with full systems of leaves, stems, and roots. They also include a few of their close relatives, often called *bryophytes*, of which mosses and liverworts are the most common.

All of these plants have eukaryotic cells with cell walls composed of cellulose, and most obtain their energy through photosynthesis, using light and carbon dioxide to synthesize food. About three hundred plant species do not photosynthesize but are parasites on other species of photosynthetic plants. Plants are distinguished from green algae, which represent a mode of photosynthetic life similar to the kind modern plants are believed to have evolved from, by having specialized reproductive organs protected by non-reproductive tissues.

*Dicksonia antarctica*, a species of tree fern

Bryophytes first appeared during the early Paleozoic. They can only survive where moisture is available for significant periods, although some species are desiccation tolerant. Most species of bryophyte remain small throughout their life-cycle. This involves an alternation between two generations: a haploid stage, called the gametophyte, and a diploid stage, called the sporophyte. The sporophyte is short-lived and remains dependent on its parent gametophyte.

Vascular plants first appeared during the Silurian period, and by the Devonian had diversified and spread into many different land environments. They have a number of adaptations that allowed them to overcome the limitations of the bryophytes. These include a cuticle resistant to desiccation, and vascular tissues which transport water throughout the organism. In most the sporophyte acts as a separate individual, while the gametophyte remains small.

The first primitive seed plants, Pteridosperms (seed ferns) and Cordaites, both groups now extinct, appeared in the late Devonian and diversified through the Carboniferous, with further evolution through the Permian and Triassic periods. In these the gametophyte stage is completely reduced, and the sporophyte begins life inside an enclosure called a seed, which develops while on the parent plant, and with fertilisation by means of pollen grains. Whereas other vascular plants, such as ferns, reproduce by means of spores and so need moisture to develop, some seed plants can survive and reproduce in extremely arid conditions.

Early seed plants are referred to as gymnosperms (naked seeds), as the seed embryo is not enclosed in a protective structure at pollination, with the pollen landing directly on the embryo. Four surviving groups remain widespread now, particularly the conifers, which are dominant trees in several biomes. The angiosperms, comprising the flowering plants, were the last major group of plants to appear, emerging from within the gymnosperms during the Jurassic and diversifying rapidly during the Cretaceous. These differ in that the seed embryo (angiosperm) is enclosed, so the pollen has to grow a tube to penetrate the protective seed coat; they are the predominant group of flora in most biomes today.

## Fossils

Plant fossils include roots, wood, leaves, seeds, fruit, pollen, spores, phytoliths, and amber (the fossilized resin produced by some plants). Fossil land plants are recorded in terrestrial, lacustrine, fluvial and nearshore marine sediments. Pollen, spores and algae (dinoflagellates and acritarchs) are used for dating sedimentary rock sequences. The remains of fossil plants are not as common as fossil animals, although plant fossils are locally abundant in many regions worldwide.

The earliest fossils clearly assignable to Kingdom Plantae are fossil green algae from the Cambrian. These fossils resemble calcified multicellular members of the Dasycladales. Earlier Precambrian fossils are known which resemble single-cell green algae, but definitive identity with that group of algae is uncertain.

The oldest known fossils of embryophytes date from the Ordovician, though such fossils are fragmentary. By the Silurian, fossils of whole plants are preserved, including the lycophyte *Baragwanathia longifolia*. From the Devonian, detailed fossils of rhyniophytes have been found. Early fossils of these ancient plants show the individual cells within the plant tissue. The Devonian period also saw the

A petrified log in Petrified Forest National Park.

evolution of what many believe to be the first modern tree, *Archaeopteris*. This fern-like tree combined a woody trunk with the fronds of a fern, but produced no seeds.

The Coal measures are a major source of Paleozoic plant fossils, with many groups of plants in existence at this time. The spoil heaps of coal mines are the best places to collect; coal itself is the remains of fossilised plants, though structural detail of the plant fossils is rarely visible in coal. In the Fossil Forest at Victoria Park in Glasgow, Scotland, the stumps of *Lepidodendron* trees are found in their original growth positions.

The fossilized remains of conifer and angiosperm roots, stems and branches may be locally abundant in lake and inshore sedimentary rocks from the Mesozoic and Cenozoic eras. Sequoia and its allies, magnolia, oak, and palms are often found.

Petrified wood is common in some parts of the world, and is most frequently found in arid or desert areas where it is more readily exposed by erosion. Petrified wood is often heavily silicified (the organic material replaced by silicon dioxide), and the impregnated tissue is often preserved in fine detail. Such specimens may be cut and polished using lapidary equipment. Fossil forests of petrified wood have been found in all continents.

Fossils of seed ferns such as *Glossopteris* are widely distributed throughout several continents of the Southern Hemisphere, a fact that gave support to Alfred Wegener's early ideas regarding Continental drift theory.

## Structure, growth, and development

Further information: Plant morphology

Most of the solid material in a plant is taken from the atmosphere. Through a process known as photosynthesis, most plants use the energy in sunlight to convert carbon dioxide from the atmosphere, plus water, into simple sugars. Parasitic plants, on the other hand, use the resources of its host to grow. These sugars are then used as building blocks and form the main structural component of the plant. Chlorophyll, a green-colored, magnesium-containing pigment is essential to this process; it is generally present in plant leaves, and often in other plant parts as well.

Plants usually rely on soil primarily for support and water (in quantitative terms), but also obtain compounds of nitrogen, phosphorus, and other crucial elemental nutrients. Epiphytic and lithophytic plants often depend on rainwater or other sources for nutrients and carnivorous plants supplement their nutrient requirements with insect prey that they capture. For the majority of plants to grow successfully they also require oxygen in the atmosphere and around their roots for respiration. However, some plants grow as submerged aquatics, using oxygen dissolved in the surrounding water, and a few specialized vascular plants, such as mangroves, can grow with their roots in anoxic conditions.

## Factors affecting growth

The genotype of a plant affects its growth. For example, selected varieties of wheat grow rapidly, maturing within 110 days, whereas others, in the same environmental conditions, grow more slowly and mature within 155 days.[32]

Growth is also determined by environmental factors, such as temperature, available water, available light, and available nutrients in the soil. Any change in the availability of these external conditions will be reflected in the plants growth.

Biotic factors are also capable of affecting plant growth. Plants compete with other plants for space, water, light and nutrients. Plants can be so crowded that no single individual produces normal growth, causing etiolation and chlorosis. Optimal plant growth can be hampered by grazing animals, suboptimal soil composition, lack of mycorrhizal fungi, and attacks by insects or plant diseases, including those caused by bacteria, fungi, viruses, and nematodes.[32]

Simple plants like algae may have short life spans as individuals, but their populations are commonly seasonal. Other plants may be organized according to their seasonal growth pattern: annual plants live and reproduce within one growing season, biennial plants live for two growing seasons and usually reproduce in second year, and perennial plants live for many growing seasons and continue to reproduce once they are mature. These designations often depend on climate and other environmental factors; plants that are annual in alpine or temperate regions can be biennial or perennial in warmer climates. Among the vascular plants, perennials include both evergreens that keep their leaves the entire year, and deciduous plants which lose their leaves for some part of it. In temperate and boreal climates, they generally lose their leaves during the winter; many tropical plants lose their leaves during the dry season.

The leaf is usually the primary site of photosynthesis in plants.

There is no photosynthesis in deciduous leaves in autumn.

The growth rate of plants is extremely variable. Some mosses grow less than 0.001 millimeters per hour (mm/h), while most trees grow 0.025-0.250 mm/h. Some climbing species, such as kudzu, which do not need to produce thick supportive tissue, may grow up to 12.5 mm/h.

Plants protect themselves from frost and dehydration stress with antifreeze proteins, heat-shock proteins and sugars (sucrose is common). LEA (Late Embryogenesis Abundant) protein expression is induced by stresses and protects other proteins from aggregation as a result of desiccation and freezing.[33]

Dried dead plants

## Plant cell

Plant cells are typically distinguished by their large water-filled central vacuole, chloroplasts, and rigid cell walls that are made up of cellulose, hemicellulose, and pectin. Cell division is also characterized by the development of a phragmoplast for the construction of a cell plate in the late stages of cytokinesis. Just as in animals, plant cells differentiate and develop into multiple cell types. Totipotent meristematic cells can differentiate into vascular, storage, protective (e.g. epidermal layer), or reproductive tissues, with more primitive plants lacking some tissue types.[34]

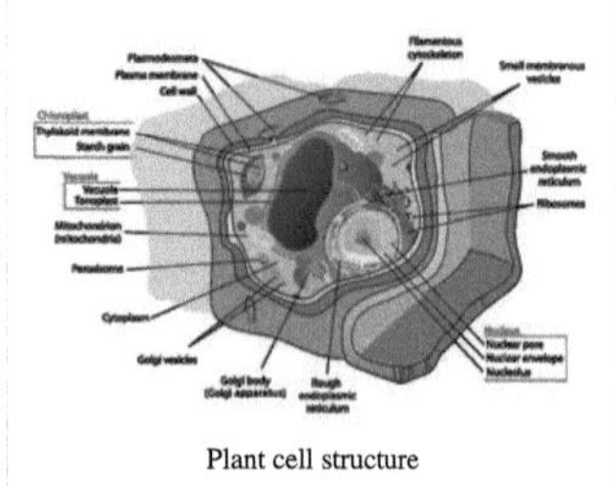
Plant cell structure

# Physiology

## Photosynthesis

Plants are photosynthetic, which means that they manufacture their own food molecules using energy obtained from light. The primary mechanism plants have for capturing light energy is the pigment chlorophyll. All green plants contain two forms of chlorophyll, chlorophyll *a* and chlorophyll *b*. The latter of these pigments is not found in red or brown algae.

## Immune system

By means of cells that behave like nerves, plants receive and distribute within their systems information about incident light intensity and quality. Incident light which stimulates a chemical reaction in one leaf, will cause a chain reaction of signals to the entire plant via a type of cell termed a *bundle sheath cell*. Researchers from the Warsaw University of Life Sciences in Poland, found that plants have a specific memory for varying light conditions which prepares their immune systems against seasonal pathogens.[35] Plants use pattern-recognition receptors to recognize conserved microbial signatures. This recognition triggers an immune response. The first plant receptors of conserved microbial signatures were identified in rice (XA21, 1995)[36] and in *Arabidopsis* (FLS2, 2000).[37] Plants also carry immune receptors that recognize highly variable pathogen effectors. These include the NBS-LRR class of proteins.

## Internal distribution

Vascular plants differ from other plants in that they transport nutrients between different parts through specialized structures, called xylem and phloem. They also have roots for taking up water and minerals. The xylem moves water and minerals from the root to the rest of the plant, and the phloem provides the roots with sugars and other nutrient produced by the leaves.[34]

# Ecology

The photosynthesis conducted by land plants and algae is the ultimate source of energy and organic material in nearly all ecosystems. Photosynthesis radically changed the composition of the early Earth's atmosphere, which as a result is now 21% oxygen. Animals and most other organisms are aerobic, relying on oxygen; those that do not are confined to relatively rare anaerobic environments. Plants are the primary producers in most terrestrial ecosystems and form the basis of the food web in those ecosystems. Many animals rely on plants for shelter as well as oxygen and food.

Land plants are key components of the water cycle and several other biogeochemical cycles. Some plants have coevolved with nitrogen fixing bacteria, making plants an important part of the nitrogen cycle. Plant roots play an essential role in soil development and prevention of soil erosion.

## Distribution

Plants are distributed worldwide in varying numbers. While they inhabit a multitude of biomes and ecoregions, few can be found beyond the tundras at the northernmost regions of continental shelves. At the southern extremes, plants have adapted tenaciously to the prevailing conditions. (See Antarctic flora.)

Plants are often the dominant physical and structural component of habitats where they occur. Many of the Earth's biomes are named for the type of vegetation because plants are the dominant organisms in those biomes, such as grasslands and forests.

## Ecological relationships

Numerous animals have coevolved with plants. Many animals pollinate flowers in exchange for food in the form of pollen or nectar. Many animals disperse seeds, often by eating fruit and passing the seeds in their feces. Myrmecophytes are plants that have coevolved with ants. The plant provides a home, and sometimes food, for the ants. In exchange, the ants defend the plant from herbivores and sometimes competing plants. Ant wastes provide organic fertilizer.

The majority of plant species have various kinds of fungi associated with their root systems in a kind of mutualistic symbiosis known as mycorrhiza. The fungi help the plants gain water and mineral nutrients from the soil, while the plant gives the fungi carbohydrates manufactured in photosynthesis. Some plants serve as homes for endophytic fungi that protect the plant from herbivores by producing toxins. The fungal endophyte, *Neotyphodium coenophialum*, in tall fescue (*Festuca arundinacea*) does tremendous economic damage to the cattle industry in the U.S.

The Venus flytrap, a species of carnivorous plant.

Various forms of parasitism are also fairly common among plants, from the semi-parasitic mistletoe that merely takes some nutrients from its host, but still has photosynthetic leaves, to the fully parasitic broomrape and toothwort that acquire all their nutrients through connections to the roots of other plants, and so have no chlorophyll. Some plants, known as myco-heterotrophs, parasitize mycorrhizal fungi, and hence act as epiparasites on other plants.

Many plants are epiphytes, meaning they grow on other plants, usually trees, without parasitizing them. Epiphytes may indirectly harm their host plant by intercepting mineral nutrients and light that the host would otherwise receive. The weight of large numbers of epiphytes may break tree limbs. Hemiepiphytes like the strangler fig begin as epiphytes but eventually set their own roots and overpower and kill their host. Many orchids, bromeliads, ferns and mosses often grow as epiphytes. Bromeliad epiphytes accumulate water in leaf axils to form phytotelmata, complex aquatic food webs.[38]

Approximately 630 plants are carnivorous, such as the Venus Flytrap (*Dionaea muscipula*) and sundew (*Drosera* species). They trap small animals and digest them to obtain mineral nutrients, especially nitrogen and phosphorus.[39]

# Importance

The study of plant uses by people is termed economic botany or ethnobotany; some consider economic botany to focus on modern cultivated plants, while ethnobotany focuses on indigenous plants cultivated and used by native peoples. Human cultivation of plants is part of agriculture, which is the basis of human civilization. Plant agriculture is subdivided into agronomy, horticulture and forestry.

## Food

Much of human nutrition depends on land plants, either directly or indirectly.

Human nutrition depends to a large extent on cereals, especially maize (or corn), wheat and rice. Other staple crops include potato, cassava, and legumes. Human food also includes vegetables, spices, and certain fruits, nuts, herbs, and edible flowers.

Beverages produced from plants include coffee, tea, wine, beer and alcohol.

Sugar is obtained mainly from sugar cane and sugar beet.

Cooking oils and margarine come from maize, soybean, rapeseed, safflower, sunflower, olive and others.

Food additives include gum arabic, guar gum, locust bean gum, starch and pectin.

Livestock animals including cows, pigs, sheep, and goats are all herbivores; and feed primarily or entirely on cereal plants, particularly grasses.

Potato plant. Potatoes spread to the rest of the world after European contact with the Americas in the late 15th and early 16th centuries and have since become an important field crop.

## Nonfood products

Wood is used for buildings, furniture, paper, cardboard, musical instruments and sports equipment. Cloth is often made from cotton, flax or synthetic fibers derived from cellulose, such as rayon and acetate. Renewable fuels from plants include firewood, peat and many other biofuels. Coal and petroleum are fossil fuels derived from plants. Medicines derived from plants include aspirin, taxol, morphine,

Timber in storage for later processing at a sawmill.

quinine, reserpine, colchicine, digitalis and vincristine. There are hundreds of herbal supplements such as ginkgo, Echinacea, feverfew, and Saint John's wort. Pesticides derived from plants include nicotine, rotenone, strychnine and pyrethrins. Drugs obtained from plants include opium, cocaine and marijuana. Poisons from plants include ricin, hemlock and curare. Plants are the source of many natural products such as fibers, essential oils, natural dyes, pigments, waxes, tannins, latex, gums, resins, alkaloids, amber and cork. Products derived from plants include soaps, paints, shampoos, perfumes, cosmetics, turpentine, rubber, varnish, lubricants, linoleum, plastics, inks, chewing gum and hemp rope. Plants are also a primary source of basic chemicals for the industrial synthesis of a vast array of organic chemicals. These chemicals are used in a vast variety of studies and experiments.

A section of a Yew branch showing 27 annual growth rings, pale sapwood and dark heartwood, and pith (centre dark spot). The dark radial lines are longitudinal sections of small branches which became included by growth of the tree.

## Aesthetic uses

Thousands of plant species are cultivated for aesthetic purposes as well as to provide shade, modify temperatures, reduce wind, abate noise, provide privacy, and prevent soil erosion. People use cut flowers, dried flowers and houseplants indoors or in greenhouses. In outdoor gardens, lawn grasses, shade trees, ornamental trees, shrubs, vines, herbaceous perennials and bedding plants are used. Images of plants are often used in art, architecture, humor, language, and photography and on textiles, money, stamps, flags and coats of arms. Living plant art forms include topiary, bonsai, ikebana and espalier. Ornamental plants have sometimes changed the course of history, as in tulipomania. Plants are the basis of a multi-billion dollar per year tourism industry which includes travel to arboretums, botanical gardens, historic gardens, national parks, tulip festivals, rainforests, forests with colorful autumn leaves and the National Cherry Blossom Festival. Venus Flytrap, sensitive plant and resurrection plant are examples of plants sold as novelties.

## Scientific and cultural uses

Tree rings are an important method of dating in archeology and serve as a record of past climates. Basic biological research has often been done with plants, such as the pea plants used to derive Gregor Mendel's laws of genetics. Space stations or space colonies may one day rely on plants for life support. Plants are used as national and state emblems, including state trees and state flowers. Ancient trees are revered and many are famous. Numerous world records are held by plants. Plants are often used as memorials, gifts and to mark special occasions such as births, deaths, weddings and holidays. Plants figure prominently in mythology, religion and literature. The field of ethnobotany studies plant use by indigenous cultures which helps to conserve endangered species as well as discover new medicinal plants. Gardening is the most popular leisure activity in the U.S. Working with plants or horticulture therapy is beneficial for rehabilitating people with disabilities. Certain plants contain psychotropic chemicals which are extracted and ingested, including tobacco, cannabis (marijuana), and opium.

## Negative effects

Weeds are plants that grow where people do not want them. People have spread plants beyond their native ranges and some of these introduced plants become invasive, damaging existing ecosystems by displacing native species. Invasive plants cause billions of dollars in crop losses annually by displacing crop plants, they increase the cost of production and the use of chemical means to control them affects the environment.

Plants may cause harm to animals, including people. Plants that produce windblown pollen invoke allergic reactions in people who suffer from hay fever. A wide variety of plants are poisonous. Toxalbumins are plant poisons fatal to most mammals and act as a serious deterrent to consumption. Several plants cause skin irritations when touched, such as poison ivy. Certain plants contain psychotropic chemicals, which are extracted and ingested or smoked, including tobacco, cannabis (marijuana), cocaine and opium. Smoking causes damage to health or even death, while some drugs may also be harmful or fatal to people.[40][41] Both illegal and legal drugs derived from plants may have negative effects on the economy, affecting worker productivity and law enforcement costs.[42][43] Some plants cause allergic reactions when ingested, while other plants cause food intolerances that negatively affect health.

# References


[1]  Haeckel G (1866). *Generale Morphologie der Organismen*. Berlin: Verlag von Georg Reimer. pp. vol.1: i–xxxii, 1–574, pls I–II; vol. 2: i–clx, 1–462, pls I–VIII.

[2]  http://www.iucnredlist.org/documents/summarystatistics/2010_1RL_Stats_Table_1.pdf

[3]  "plant[2 (http://www.merriam-webster.com/dictionary/plant[2]) - Definition from the Merriam-Webster Online Dictionary"]. . Retrieved 2009-03-25.

[4]  "plant (life form) -- Britannica Online Encyclopedia" (http://www.britannica.com/EBchecked/topic/463192/plant). . Retrieved 2009-03-25.

[5]  Based on Rogozin, I.B.; Basu, M.K.; Csürös, M. & Koonin, E.V. (2009), "Analysis of Rare Genomic Changes Does Not Support the Unikont–Bikont Phylogeny and Suggests Cyanobacterial Symbiosis as the Point of Primary Radiation of Eukaryotes", *Genome Biology and Evolution* **1**: 99–113, doi:10.1093/gbe/evp011, PMC 2817406, PMID 20333181 and Becker, B. & Marin, B. (2009), "Streptophyte algae and the origin of embryophytes", *Annals of Botany* **103** (7): 999–1004, doi:10.1093/aob/mcp044, PMC 2707909, PMID 19273476; see also the slightly different cladogram in Lewis, Louise A. & McCourt, R.M. (2004), "Green algae and the origin of land plants", *Am. J. Bot.* **91** (10): 1535–1556, doi:10.3732/ajb.91.10.1535, PMID 21652308.

[6]  Margulis, L. (1974). "Five-kingdom classification and the origin and evolution of cells". *Evolutionary Biology* **7**: 45–78.

[7]  Raven, Peter H., Ray F. Evert, & Susan E. Eichhorn, 2005. *Biology of Plants*, 7th edition. (New York: W. H. Freeman and Company). ISBN 0-7167-1007-2.

[8]  Lewis, Louise A. & McCourt, R.M. (2004), "Green algae and the origin of land plants", *Am. J. Bot.* **91** (10): 1535–1556, doi:10.3732/ajb.91.10.1535, PMID 21652308

[9]  Becker, B. & Marin, B. (2009), "Streptophyte algae and the origin of embryophytes", *Annals of Botany* **103** (7): 999–1004, doi:10.1093/aob/mcp044, PMC 2707909, PMID 19273476

[10]  Guiry, M.D. & Guiry, G.M. (2007). "Phylum: Chlorophyta taxonomy browser" (http://www.algaebase.org/browse/taxonomy/?id=4307). *AlgaeBase version 4.2* World-wide electronic publication, National University of Ireland, Galway. . Retrieved 2007-09-23.

[11]  Van den Hoek, C., D. G. Mann, & H. M. Jahns, 1995. *Algae: An Introduction to Phycology*. pages 343, 350, 392, 413, 425, 439, & 448 (Cambridge: Cambridge University Press). ISBN 0-521-30419-9

[12]  Guiry, M.D. & Guiry, G.M. (2011), *AlgaeBase : Chlorophyta* (http://www.algaebase.org/browse/taxonomy/?searching=true& gettaxon=Chlorophyta), World-wide electronic publication, National University of Ireland, Galway, , retrieved 2011-07-26

[13]  Guiry, M.D. & Guiry, G.M. (2011), *AlgaeBase : Charophyta* (http://www.algaebase.org/browse/taxonomy/?searching=true& gettaxon=Charophyta), World-wide electronic publication, National University of Ireland, Galway, , retrieved 2011-07-26

[14]  Van den Hoek, C., D. G. Mann, & H. M. Jahns, 1995. *Algae: An Introduction to Phycology*. pages 457, 463, & 476. (Cambridge: Cambridge University Press). ISBN 0-521-30419-9

[15]  Crandall-Stotler, Barbara. & Stotler, Raymond E., 2000. "Morphology and classification of the Marchantiophyta". page 21 *in* A. Jonathan Shaw & Bernard Goffinet (Eds.), *Bryophyte Biology*. (Cambridge: Cambridge University Press). ISBN 0-521-66097-1

[16]  Schuster, Rudolf M., *The Hepaticae and Anthocerotae of North America*, volume VI, pages 712-713. (Chicago: Field Museum of Natural History, 1992). ISBN 0-914868-21-7.

[17]  Goffinet, Bernard; William R. Buck (2004). "Systematics of the Bryophyta (Mosses): From molecules to a revised classification". *Monographs in Systematic Botany* (Missouri Botanical Garden Press) **98**: 205–239.

[18]  Gifford, Ernest M. & Adriance S. Foster, 1988. *Morphology and Evolution of Vascular Plants*, 3rd edition, page 358. (New York: W. H. Freeman and Company). ISBN 0-7167-1946-0.

[19] Taylor, Thomas N. & Edith L. Taylor, 1993. *The Biology and Evolution of Fossil Plants*, page 636. (New Jersey: Prentice-Hall). ISBN 0-13-651589-4.

[20] International Union for Conservation of Nature and Natural Resources, 2006. *IUCN Red List of Threatened Species:Summary Statistics* (http://www.iucnredlist.org/)

[21] http://toolserver.org/~verisimilus/Timeline/Timeline.php?Ma=1200

[22] http://toolserver.org/~verisimilus/Timeline/Timeline.php?Ma=450

[23] "The oldest fossils reveal evolution of non-vascular plants by the middle to late Ordovician Period (~450-440 m.y.a.) on the basis of fossil spores" Transition of plants to land (http://www.clas.ufl.edu/users/pciesiel/gly3150/plant.html)

[24] "The apparent dominance of eukaryotes in non-marine settings by 1 Gyr ago indicates that eukaryotic evolution on land may have commenced far earlier than previously thought." Earth's earliest non-marine eukaryotes (http://www.nature.com/nature/journal/vaop/ncurrent/full/nature09943.html)

[25] http://toolserver.org/~verisimilus/Timeline/Timeline.php?Ma=420

[26] Rothwell, G. W.; Scheckler, S. E.; Gillespie, W. H. (1989). "*Elkinsia* gen. nov., a Late Devonian gymnosperm with cupulate ovules". *Botanical Gazette* **150** (2): 170–189. doi:10.1086/337763.

[27] http://toolserver.org/~verisimilus/Timeline/Timeline.php?Ma=200

[28] http://toolserver.org/~verisimilus/Timeline/Timeline.php?Ma=40

[29] http://toolserver.org/~verisimilus/Timeline/Timeline.php?Ma=10

[30] Kenrick, Paul & Peter R. Crane. 1997. *The Origin and Early Diversification of Land Plants: A Cladistic Study*. (Washington, D.C.: Smithsonian Institution Press). ISBN 1-56098-730-8.

[31] Smith Alan R., Pryer Kathleen M., Schuettpelz E., Korall P., Schneider H., Wolf Paul G. (2006). "A classification for extant ferns" (http://www.pryerlab.net/publication/fichier749.pdf) (PDF). *Taxon* **55** (3): 705–731. doi:10.2307/25065646. .

[32] Robbins, W.W., Weier, T.E., *et al.*, *Botany:Plant Science*, 3rd edition , Wiley International, New York, 1965.

[33] Goyal, K., Walton, L. J., & Tunnacliffe, A. (2005). "LEA proteins prevent protein aggregation due to water stress" (http://www.webcitation.org/5il9QhYT0). *Biochemical Journal* **388** (Part 1): 151–157. doi:10.1042/BJ20041931. PMC 1186703. PMID 15631617. Archived from the original (http://www.biochemj.org/bj/388/0151/bj3880151.htm) on 2009-08-03. .

[34] Campbell, Reece, *Biology*, 7th edition, Pearson/Benjamin Cummings, 2005.

[35] BBC Report (http://www.bbc.co.uk/news/10598926)

[36] Song, W.Y. et al. (1995). "A receptor kinase-like protein encoded by the rice disease resistance gene, XA21". *Science* **270** (5243): 1804–1806. doi:10.1126/science.270.5243.1804. PMID 8525370.

[37] Gomez-Gomez, L. et al. (2000). "FLS2: an LRR receptor-like kinase involved in the perception of the bacterial elicitor flagellin in *Arabidopsis*". *Molecular Cell* **5** (6): 1003–1011. doi:10.1016/S1097-2765(00)80265-8. PMID 10911994.

[38] Howard Frank, Bromeliad Phytotelmata (http://entomology.ifas.ufl.edu/frank/bromeliadbiota/bromfit.htm), October 2000

[39] Barthlott, W., S. Porembski, R. Seine, and I. Theisen. 2007. *The Curious World of Carnivorous Plants: A Comprehensive Guide to Their Biology and Cultivation*. Timber Press: Portland, Oregon.

[40] "cocaine/crack" (http://www.urban75.com/Drugs/drugcoke.html). .

[41] "Deaths related to cocaine" (http://ar2005.emcdda.europa.eu/en/page050-en.html). .

[42] "Illegal drugs drain $160 billion a year from American economy" (http://web.archive.org/web/20080215071055/http://www.whitehousedrugpolicy.gov/NEWS/press02/012302.html). Archived from the original (http://www.whitehousedrugpolicy.gov/NEWS/press02/012302.html) on 2008-02-15. .

[43] "The social cost of illegal drug consumption in Spain" (http://www.ingentaconnect.com/content/bsc/add/2002/00000097/00000009/art00012). .

## Further reading

General

- Evans, L. T. (1998). *Feeding the Ten Billion - Plants and Population Growth*. Cambridge University Press. Paperback, 247 pages. ISBN 0-521-64685-5.

- Kenrick, Paul & Crane, Peter R. (1997). *The Origin and Early Diversification of Land Plants: A Cladistic Study*. Washington, D. C.: Smithsonian Institution Press. ISBN 1-56098-730-8.

- Raven, Peter H., Evert, Ray F., & Eichhorn, Susan E. (2005). *Biology of Plants* (7th ed.). New York: W. H. Freeman and Company. ISBN 0-7167-1007-2.

- Taylor, Thomas N. & Taylor, Edith L. (1993). *The Biology and Evolution of Fossil Plants*. Englewood Cliffs, NJ: Prentice Hall. ISBN 0-13-651589-4.

- Trewavas A (2003). "Aspects of Plant Intelligence" (http://aob.oxfordjournals.org/cgi/content/full/92/1/1). *Annals of Botany* **92**: 1–20.

Species estimates and counts

- International Union for Conservation of Nature and Natural Resources (IUCN) Species Survival Commission (2004). IUCN Red List (http://www.iucnredlist.org/).
- Prance G. T. (2001). "Discovering the Plant World". *Taxon* **50**: 345–359.

## External links

- *Plant* (http://www.eol.org/pages/281) at the Encyclopedia of Life
- Chaw, S.-M. et al. (1997). "Molecular Phylogeny of Extant Gymnosperms and Seed Plant Evolution: Analysis of Nuclear 18s rRNA Sequences" (http://mbe.library.arizona.edu/data/1997/1401/7chaw.pdf). *Molec. Biol. Evol.* **14** (1): 56–68. PMID 9000754.
- Index Nominum Algarum (http://ucjeps.berkeley.edu/INA.html)
- Interactive Cronquist classification (http://florabase.calm.wa.gov.au/phylogeny/cronq88.html)
- Plant Photo Gallery of Japan (http://www.alpine-plants-jp.com/art/index_photo2b.htm) - Flavon's Wild herb and Alpine plants
- Plant Picture Gallery (http://www.pflanzenliebe.de/)
- Plant Resources of Tropical Africa (http://www.prota.org/uk/About+Prota/)
- www.prota.org - PROTA's mission (http://database.prota.org/search.htm)
- Tree of Life (http://tolweb.org/Green_plants)

Botanical and vegetation databases

- African Plants Initiative database (http://www.aluka.org/action/doBrowse?sa=1&sa_sel=)
- Australia (http://www.anbg.gov.au/cpbr/databases/)
- Chilean plants at *Chilebosque* (http://www.chilebosque.cl/)
- Dave's garden (http://davesgarden.com/pf/) plenty of information mostly about garden plants
- e-Floras (Flora of China, Flora of North America and others) (http://www.efloras.org/index.aspx)
- Flora Europaea (http://rbg-web2.rbge.org.uk/FE/fe.html)
- Flora of Central Europe (http://www.floraweb.de/) (German)
- Flora of North America (http://www.efloras.org/flora_page.aspx?flora_id=1)
- List of Japanese Wild Plants Online (http://www.alpine-plants-jp.com/botanical_name/ list_of_japanese_wild_plants_abelia_buxus.htm)
- Meet the Plants-National Tropical Botanical Garden (http://www.ntbg.org/plants/choose_a_plant.php)
- Lady Bird Johnson Wildflower Center - Native Plant Information Network at University of Texas, Austin (http://www.wildflower.org/)
- The Plant List (http://www.theplantlist.org/)
- United States Department of Agriculture (http://plants.usda.gov/) not limited to continental US species

# Article Sources and Contributors

**Malva parviflora**  *Source*: http://en.wikipedia.org/w/index.php?title=Malva_parviflora  *Contributors*: Atubeileh, Curtis Clark, Dtrebbien, Dysmorodrepanis, Elie plus, Extra bases, Hesperian, IceCreamAntisocial, Lotje, Makki98, Melburnian, Michael Bailes, عمرو بن كلثوم, 1 anonymous edits

**North Africa**  *Source*: http://en.wikipedia.org/w/index.php?title=North_Africa  *Contributors*: !ComputerAlert!, 195.149.37.xxx, 24.251.118.xxx, 44 forty four, A Jalil, A.Khalil, A2e4c97, ABF, ARC Gritt, Abhijitsathe, Abjiklam, Acalamari, Acerperi, Ackees, Adam78, Adam7davies, Adel the king, Ahoerstemeier, Ahuskay, Al-Andalus, Alamawi, Alansohn, Alexf, Almountasir, Alsandro, Amazigh Man, Ameno, Amog, Andres, Ann Stouter, Antdawg22, Antzervos, Anwar saadat, Arab Hafez, ArielGold, Arnie Gov, Arre, Atamata, BD2412, BanyanTree, Beland, Belovedfreak, Beyond My Ken, Bggoldie, Big Adamsky, Black Falcon, Blanchardb, Bluerasberry, Bobblehead, Booboo7538, Boobylover, Bouha, Bradv, BrendanRyan, BrianCumminger, Buaidh, Bucketsofg, C0nanPayne, Calabe1992, CallMeLee, Caltas, CanadianWikipedian2008, CardinalDan, Caribbean H.Q., Causteau, Cbustapeck, Chan Yin Keen, Chase me ladies, I'm the Cavalry, Cholmes75, Chongkian, ChrisHamburg, ChrisSk8, Chrism, Christopher Parham, City of Tragedy, Ckatz, ClamDip, Cluckbang, Cncs wikipedia, Collounsbury, Connormah, Conversion script, Corticopia, Creidieki, CrimeCentral, Cst17, Cureden, DARTH SIDIOUS 2, DB, DMacks, Dabbler, Dance21c, DarkFalls, Darkwind, Dbachmann, DeadEyeArrow, Debresser, Deconstructhis, Deeceevoice, Deerlike, Delivernews, Den fjättrade ankan, Dialectric, Diannaa, Dimadick, Discopinster, Dkusic, DocWatson42, Dogru144, Donutcity, Dreadstar, Dreamequal88, Drmaik, Drmies, Drpickem, Dswislow, Duncan, DéRahier, E Pluribus Anthony, E2eamon, Eagleamn, Egyptian lion, El C, El Moro, Eskandarany, Everyking, Ewebb49, Ezeu, Fastily, Fayaonline, FayssalF, Felliax08, Ffx1619, Fieldday-sunday, Floquenbeam, FrFintonStack, Frank Scipio, Fratrep, Frawetfva, Full Shunyata, Funnyfarmofdoom, GHe, Gailtb, Gameboyguy13, Gamefreakxx, Garyzx, Garzo, Gentgeen, Geographyprof, Glacialfox, Gnowor, Graham87, Greyshark09, Grutness, Haham hanuka, Halaqah, Haleth, Hasslich, Hattar393, Hazeni, Heimstern, Heraldicos, Hermitstudy, Heron, HexaChord, Hispanic90210, Hmains, Hotlorp, Hunnjazal, IRP, IW.HG, Ibagli, Ibrahim@ancientsudan.org, Imnotminkus, Indon, Instinct, Iritakamas, Its snowing in East Asia, ItsZippy, J.delanoy, JForget, JackyR, Jadtnr1, JamesBWatson, Jatebirds, Jayjg, Jerry, Jespinos, Jewbask, Jigglyfidders, JimWae, Jimtaip, Jmac800, Jncraton, Joefromrandb, Jojit fb, Joy, Jusdafax, Just James, Kafka1, Kanjaristan, Keilana, Kellen548, Khalid hassani, Khalidmn, Khoikhoi, Kingpin13, Klilidiplomus, Koavf, Kralizec!, Kudzu1, Kwamikagami, Kylerhaha, L Kensington, LailaKes, LambaJan, Leonard G., Lesliestng, Levineps, Lexina Glow, LilHelpa, LindsayH, Livajo, Logan, Lonewolf BC, Lottamiata, Lucidity, MER-C, Magister Mathematicae, Malcolm Farmer, Marcotime, Mardus, Mariam83, Marianian, Mark91, Marnanel, Materialscientist, MauriManya, Maureen, Maximaximax, Mazigh-berber, McSly, Mcorazao, Mic, Micaburn, Mightymights, Mihxil, MikeVitale, Mild Bill Hiccup, Minimac, Miros 0571, Missionary, Mit32, Mmacia812, Montrealais, Moroccanmaryam, Moverton, Mrrussianhunter, MsDivagin, Msruzicka, Mttll, Muijz, Mustafaa, Nakon, Natalina95, NerdyScienceDude, Netknowle, Neurolysis, Neutrality, Nicoletta8383, Njaelkies Lea, Nuttycoconut, Ogress, Olivier, Omar-Toons, Omicronpersei8, Orestek, Oscarthecat, Ouedbirdwatcher, Oxymoron83, Parthava, Patrick, Pdeitiker, PedroPVZ, Pekayer11, Philip Trueman, Phoenix B 1of3, Piano non troppo, Picaroon, Pigman, Pinethicket, Plumbstar, Potosino, Povertypop, Pras, Prodego, Prolog, ProtoFire, PseudoSudo, Quadell, Quincy2010, R'n'B, RainbowOfLight, RandomP, Randy Johnston, Raven in Orbit, Rdsmith4, Reaper Eternal, Rediahs, ReinforcedReinforcements, Rentaferret, RexNL, Richard D. LeCour, Rjwilmsi, Rojypala, Romanm, Romanskolduns, Rowerlali, Roylee, Rumping, ST47, Sabbe, Sango123, Saranghae honey, Saturn6, SchfiftyThree, Seb az86556, Sen agri, Shadowjams, Shamir1, Shawn K. Quinn, Sibi antony, Silver Edge, Simesa, Simon J Kissane, SimonP, SineWave, Sir Nicholas de Mimsy-Porpington, Skafis, Skäpperöd, Slawojarek, Snoyes, Songalulu, Sophie, Sophus Bie, SteveLo, Sunshine4921, Supercoop, Sverdrup, Tachfin, Taharqa, Tataryn77, Tbone, Tellyaddict, Tgeairn, The Illusive Man, The Man in Question, The Thing That Should Not Be, The wub, TheOtherJesse, Theccy, Theflag21, ThinkBlue, ThinkEnemies, Tiddly Tom, Tide rolls, Tobby72, Tobyc75, Tom Sauce, Tommy2010, Tpbradbury, Travelbird, Trevor MacInnis, Triwbe, Ttwaring, Tussna, TwinsFan48, Twsx, Txomin, Ursutraide, Varlaam, Vegaswikian, Velella, Viktor-viking, VirtualDelight, Vlma111, Voxpuppet, Vuong Ngan Ha, Warofdreams, Washdivad, Washmuaria, Wavelength, Wayne Slam, Wertoy36, WhisperToMe, Wikih101, Wikipelli, Wikiscribe, Willdude123, Wing Nut, Wiplar, Wolfling, Woohookitty, Wsese2, Www06035, XPTO, Yobaranut, Yom, Yonghokim, Zaza8675, Zenithfel, Zerida, الجكبر, ترجمان 05, 925 anonymous edits

**Malva**  *Source*: http://en.wikipedia.org/w/index.php?title=Malva  *Contributors*: AThing, Agyle, Anlace, Bige1977, Bogdangiusca, Brighterorange, CalicoCatLover, CarolSpears, Codrinb, Cuaxdon, Djlayton4, DocWatson42, Droll, Ebizur, Emijrp, Fangjian, Fastifex, Gilabrand, Graham87, Guypadfield, Hairwizard91, Hesperian, JMBryant, JaGa, Kingdon, Lavateraguy, LuoShengli, MPF, Makki98, Matthewcgirling, Melburnian, Mgoodyear, Middayexpress, Nugunya, Orionist, Pekinensis, Professorzed, R'n'B, Richard Barlow, Rkitko, Shao, TDogg310, XJamRastafire, Yworo, مانی, 35 anonymous edits

**Malvaceae**  *Source*: http://en.wikipedia.org/w/index.php?title=Malvaceae  *Contributors*: Aelwyn, Andrew Dalby, Atubeileh, Azalea pomp, Azhyd, Berton, Bomac, Brya, CanisRufus, Caricologist, DanielCD, Dysmorodrepanis, Eleassar, Eliz81, Eric Kvaalen, Extra999, Fratrep, Gdr, GeoGreg, Glenn, Grafen, Haim Berman, Hardyplants, Hesperian, Iorsh, Jaknouse, JamesAM, Jer ome, Jlittlet, Jmgarg1, JoJan, Jtuba, Kember, Kevmin, Krasanen, Lavateraguy, Liné1, Logan, Lotje, MPF, Mani1, Marshman, Melaen, Melburnian, MightyWarrior, Mike Dallwitz, Moflg, MrDarwin, Mukogodo, Nadiatalent, Nazca Turtlehead, Nk, Nono64, Nukeless, Nurg, Obsidian Soul, Od Mishehu, Pcb21, Peter Greenwell, Philiptdotcom, Piano non troppo, Pion, Renato Caniatti, Reporein, RichiH, Rkitko, Schzmo, Stan Shebs, TeunSpaans, UtherSRG, Vald, Varlaam, Vbs, Vishnava, Vuong Ngan Ha, WormRunner, YYOOYY, 48 anonymous edits

**Byttnerioideae**  *Source*: http://en.wikipedia.org/w/index.php?title=Byttnerioideae  *Contributors*: Bff, Kontos, TDogg310

**Malvales**  *Source*: http://en.wikipedia.org/w/index.php?title=Malvales  *Contributors*: Altenmann, Azhyd, Bo1234567890, Borgx, Brya, CambridgeBayWeather, Conversion script, Eclecticology, Gabbe, Grafen, Hesperian, Indu Singh, Iorsh, Jaknouse, Jmgarg1, Josh Grosse, Julia Rossi, Lavateraguy, Llywrch, Looxix, Lord Voldemort, MPF, Marshman, Martpol, Maxima m, Mike Dallwitz, MrDarwin, Naddy, Nk, Nurg, Pekinensis, Rjwilmsi, Rkitko, Stemonitis, Sylverfysh, TeunSpaans, Tom Radulovich, Unyoyega, Vicpeters, Vina, Vuong Ngan Ha, Westerness, Wiwaxia, WooteleF, عمرو بن كلثوم, 12 anonymous edits

**Brownlowioideae**  *Source*: http://en.wikipedia.org/w/index.php?title=Brownlowioideae  *Contributors*: Dysmorodrepanis, Hesperian, JoJan, Rkitko, 1 anonymous edits

**Rosids**  *Source*: http://en.wikipedia.org/w/index.php?title=Rosids  *Contributors*: Brya, Cedders, Cephal-odd, EncycloPetey, Fanghong, Grafen, Hesperian, Hqb, Human anatomy, Jmgarg1, Kingdon, Kleopatra, Kupirijo, Lavateraguy, Mccready, Medeis, Nadiatalent, Nurg, PaleCloudedWhite, PhiLiP, Rjwilmsi, Rkitko, Sentausa, Snuffles72, Stemonitis, Summer Song, Syp, Vuong Ngan Ha, 32 anonymous edits

**Plant**  *Source*: http://en.wikipedia.org/w/index.php?title=Plant  *Contributors*: 1nic, 321mister, Abductive, Acalamari, Accurizer, Ace ETP, Adam McMaster, Adambro, Adashiel, Adenosine, Adsomvilay, Agong1, Agrihouse, Ahoerstemeier, Alan Liefting, Aleksa Lukic, Ali, Alink, Almafeta, Almog6564, Altenmann, Alwolff55, AmiDaniel, Anclation, Andre Engels, Andrew R. White, Andrewjlockley, Andrewpmk, Andris, Androstachys, Andux, Andyjsmith, Anger22, Angielaj, Anna Frodesiak, Anomalocaris, Antandrus, AnthonyBryars, AntiVan, Antonrojo, Anwar saadat, AquamarineOnion, Arcadian, Arfan, Arjun01, Arjuna316, Arkuat, Arnorian, Arsalan daudi, Art LaPella, Ashley Y, Ashmoo, Astropithicus, AtomicDragon, AuburnPilot, Aude, Autocracy, AxelBoldt, AxiomShell, Banes, BanyanTree, Barbara Shack, Barklund, Barneca, Barticus88, BaseballDetective, Bdiscoe, Bdwolverine87, Bearly541, Bearman4, Beezhive, Benbest, Bendzh, Bensaccount, Bergsten, BerneyBoy, Bewareofdog, Bibliomaniac15, Billare, Bleach926, Bobblewik, Bobianite, Bobo192, Bomac, Bongwarrior, BorgQueen, BorisTM, Borislav, Bosniak, Brainster 101, Branddobbe, Brastein, Briséis, Bubba hotep, C.Fred, CIreland, Caca, Cadiomals, Caltas, Calvin Limuel, Can't sleep, clown will eat me, CanDo, CapitalR, Cephal-odd, Chefyingi, Chickenflicker, Chipmunkdavis, Chizeng, Christian List, Ciaoriki, Circeus, Ckatz, Clemwang, Cmdrjameson, Conscious, Conversion script, Cornflake4, Corp1117, Cratylus3, Crazychinchilla, Cremepuff222, Croat Canuck, Crusadenilliteracy, Csparr, Ctbolt, Curtis Clark, CurtisLambert, Cwmhiraeth, Czj, DVD R W, Dacoutts, Daderot, Damieng, Dan East, Dan Guan, Dangerousnerd, DanielCD, Dannown, Dappawit, Darwinek, Db099221, Dcflyer, Dcoetzee, DeadEyeArrow, DeliciousT, Demmy, Deor, DerHexer, Dfrg.msc, Dinoguy2, Dipics, Directoryguru, Discospinster, Diucón, Dlloyd, Dlohcierekim, Doc glasgow, DocWatson42, Doloco, Dominus, Donald Albury, Donarreiskoffer, DougEDoug, Doyley, Dr.Bastedo, Dubaduba, Dv82matt, ESkog, Ed g2s, Edcolins, Edward321, Eleassar777, Eminem95941, Emo childnc1993, EmperorbÍma, EncycloPetey, EricG, Ettrig, Evercat, Everyking, F 22, FF2010, Fairsing, Farrahi, Femto, Filemon, Finlux, Fir0002, First Light, Floaterfluss, Fox6453, Franciepants18, FreeKresge, FreplySpang, Frogking51, Frymaster, Fumple3222, Funandtrvl, Fvasconcellos, Gabbe, Gaius Cornelius, Galoubet, Gdarin, Gdr, Geniac, Geo987, Geologyguy, Ghamer01, Giftlite, Gilliam, Gimmie, Giraffedata, Glenn, Gmunder, Gnangarra, Goatasaur, Gogo Dodo, Gop 24fan, Gotmilk954, Gracenotes, Graham87, Graminophile, Greatestrowerever, Guanaco, Guettarda, Gurch, Gveret Tered, Gwynplaine, Hadal, Haemo, Hall Monitor, Hapsiainen, Hardyplants, Hattes, Hdt83, Hdurina, Heegoop, Heimstern, Helikophis, HelloMojo, HenkvD, Henriette, HenryLi, Hephaestos, Heron, Heyitschrisss, Hi112233, Hi332211, Hmains, Hordaland, Hotty 8p, Hu12, Husond, Hyark, II MusLiM HyBRiD II, Ichormosquito, Ikanreed, Iloveplants, Infrogmation, Inner Earth, InvisibleK, IronGargoyle, Ischa1, Ixfd64, JForget, JK23, JLaTondre, JYi, Jackl, Jaknouse, Jamymc, Janetleopold, Jannex, Japs eye, Jason Leach, Jay2332, JayW, Jclerman, Jeepday, JemGage, Jemsherlock, JeremyA, Jergen, Jeronimo, Jerry, JesseGarrett, Jhagadurn, Jiddisch, JinJian, Jkyser, JoJan, Joan-of-arc, JoanneB, JoeSmack, Jolt76, Josh Grosse, Joyous!, Jpatros, Junglecat, Jurj, Juzeris, Jwanders, Jyril, Kakorot84, Karl-Henner, Karlthegreat, Karukera, Kdammers, Kesac, Ketchupabi, Kholiana joy, Kidpoker15, Kilva, Kim Bruning, Kingdon, Klemen Kocjancic, Kleopatra, Kniveswood, Kopro002, Kostisl, Kostya, Kowey, Kprwiki, Krinkle, Kungfuadam, Kuohatti, Kupirijo, Kuru, Kyleistwo, Lahiru k, Lancevortex, Lankiveil, LarryMorseDCOhio, Latitude0116, Lavateraguy, Lazulilasher, Lcgarcia, Le chien manquee, Leafyplant, Lenticel, Lesfreck, Lexor, Liface, Lights, Ligulem-s, Livestock kills, Llull, Lola lafanda, Looxix, Lord Emsworth, Lucyin, Luna Santin, Lunz2121, MER-C, MK8, MPF, MacGyverMagic, Magister Mathematicae, Magnus Manske, Maitch, Majorly, Malcolm Farmer, Man vyi, Marietta Georgia, Marrovi, MarsRover, Marshall111, Marshman, Matt Crypto, Matthew Wilson, Mav, Maxim, Mbz1, McMondongo, Mediamute, Melburnian, Menchi, Methcub, Mfatic, Mgiganteus1, Michaplot, Mike Rosoft, Million Moments, Minna Sora no Shita, Miss Madeline, Misza13, Moe Epsilon, Mofomojo, Moman, Monedula, Money4nothing, Monobi, Monolith224, Moon&Nature, Motorbikematt, Moyogo, Mr Chuckles, Mr Stephen, MrDarwin, MrFish, Msikma, Mxn, Mário e Dário, Naddy, Narayanese, Naryanlal, Natalie Erin, Natski23, NatureA16, Naveen Sankar, NawlinWiki, Nayvik, Neale Monks, Netoholic, Neverquick, NewEnglandYankee, Niluop, Ninjaboi887, Nivix, Njrfrog, Nk, Nohat, NorCal764, Northamerica1000, Nwbeeson, OMGrace, OOJaxxOo, Obli, Oblivion667, Ocatecir, Oceans and oceans, Ojaswi joshi, Oleg Alexandrov, Olivierd, Olyashok, Omicronpersei8, Onco p53, Op. Deo, Opelio, Optakeover, Osborne, Ospalh, Ost316, OwenX, Oxymoron83, P30Carl, PARA19, PKM, Paaerduag, Part Deux, Patstuart, Paul Murray, Paulbob, Pauli133, Pavel Vozenilek, Pb30, Pedroalexandrade, Pekaje, Per Ardua, Perfect Proposal, Persian Poet Gal, Peruvianllama, PetTrees, Peter G Werner, Peter coxhead, Petowl, Phaedriel, Phatvet21, Philip Trueman, PierreAbbat, Pippu d'Angelo, Plantguy, Plumbago, Poindexter Propellerhead, Polyhedron, Poopisme, Poor Yorick, Posem4012, Possum, Prodego, Pseudomonas, Pthag, PuzzletChung, Pyrospirit, Python eggs, QuartierLatin1968, R0uge, RadiantRay, RadicalBender, Radiojon, RaffiKojian, Raichu,

# Image Sources, Licenses and Contributors

**file:Malva parviflora small.jpg**  *Source*: http://en.wikipedia.org/w/index.php?title=File:Malva_parviflora_small.jpg  *License*: Creative Commons Attribution 3.0  *Contributors*: Forest & Kim Starr

**File:North Africa (orthographic projection).svg**  *Source*: http://en.wikipedia.org/w/index.php?title=File:North_Africa_(orthographic_projection).svg  *License*: Creative Commons Attribution-Sharealike 3.0  *Contributors*: Connormah

**File:Africa Koppen Map.png**  *Source*: http://en.wikipedia.org/w/index.php?title=File:Africa_Koppen_Map.png  *License*: Creative Commons Attribution-Sharealike 3.0  *Contributors*: Peel, M. C., Finlayson, B. L., and McMahon, T. A.(University of Melbourne)

**File:Flag of Algeria.svg**  *Source*: http://en.wikipedia.org/w/index.php?title=File:Flag_of_Algeria.svg  *License*: Public Domain  *Contributors*: This graphic was originaly drawn by User:SKopp.

**File:Flag of Egypt.svg**  *Source*: http://en.wikipedia.org/w/index.php?title=File:Flag_of_Egypt.svg  *License*: Public Domain  *Contributors*: Open Clip Art

**File:Flag of Libya.svg**  *Source*: http://en.wikipedia.org/w/index.php?title=File:Flag_of_Libya.svg  *License*: Public Domain  *Contributors*: Various

**File:Flag of Morocco.svg**  *Source*: http://en.wikipedia.org/w/index.php?title=File:Flag_of_Morocco.svg  *License*: Public Domain  *Contributors*: Denelson83, Zscout370

**File:Flag of South Sudan.svg**  *Source*: http://en.wikipedia.org/w/index.php?title=File:Flag_of_South_Sudan.svg  *License*: Public Domain  *Contributors*: A7x, Andrwsc, Anime Addict AA, AnonMoos, Antemister, Antonsusi, Axpde, B1mbo, Denelson83, Fred the Oyster, Fry1989, Giro720, HansenBCN, Homo lupus, Kudzu1, Maks Stirlitz, Mattes, Mnmazur, NeverDoING, Nightstallion, OAlexander, Orionist, Smooth O, SouthSudan, Sven-steffen arndt, TFCforever, Tbhotch, Zscout370, 1 anonymous edits

**File:Flag of Sudan.svg**  *Source*: http://en.wikipedia.org/w/index.php?title=File:Flag_of_Sudan.svg  *License*: Public Domain  *Contributors*: Vzb83

**File:Flag of Tunisia.svg**  *Source*: http://en.wikipedia.org/w/index.php?title=File:Flag_of_Tunisia.svg  *License*: Public Domain  *Contributors*: Alkari, AnonMoos, Avala, Bender235, Duduziq, Elina2308, Emmanuel.boutet, Flad, Fry1989, Gabbe, Juiced lemon, Klemen Kocjancic, Mattes, Meno25, Moumou82, Myself488, Neq00, Nightstallion, Reisio, Str4nd, TFCforever, Ö, Фёдор Гусляров, 9 anonymous edits

**File:Flag of the Sahrawi Arab Democratic Republic.svg**  *Source*: http://en.wikipedia.org/w/index.php?title=File:Flag_of_the_Sahrawi_Arab_Democratic_Republic.svg  *License*: Public Domain  *Contributors*: Aboyado, Anime Addict AA, AnonMoos, Avala, B1mbo, Bryan, Dbenbenn, Ed veg, Eddo, EmilJ, EugeneZelenko, Fry1989, Homo lupus, Indif, Isno, Johnonline, Juiced lemon, Ke.Pioneer, Koavf, LX, Neq00, Nightstallion, Odder, Omar-Toons, Reisio, Rocket000, Tetouancity, The Evil IP address, The White Lion, Tingo Chu, Tony Wills, Vispec, Xiquet, Zscout370, 3 anonymous edits

**Image:Septimius Severus Glyptothek Munich 357.jpg**  *Source*: http://en.wikipedia.org/w/index.php?title=File:Septimius_Severus_Glyptothek_Munich_357.jpg  *License*: Public Domain  *Contributors*: User:Bibi Saint-Pol

**File:Kairouan Mosque Courtyard.jpg**  *Source*: http://en.wikipedia.org/w/index.php?title=File:Kairouan_Mosque_Courtyard.jpg  *License*: Creative Commons Attribution-Sharealike 2.0  *Contributors*: Colin Hepburn

**file:Malva-sylvestris-20070430-1.jpg**  *Source*: http://en.wikipedia.org/w/index.php?title=File:Malva-sylvestris-20070430-1.jpg  *License*: Creative Commons Attribution-Sharealike 2.5  *Contributors*: Luis Fernández García

**File:Pavonia odorata in Talakona forest, AP W IMG 8604.jpg**  *Source*: http://en.wikipedia.org/w/index.php?title=File:Pavonia_odorata_in_Talakona_forest,_AP_W_IMG_8604.jpg  *License*: Creative Commons Attribution 3.0  *Contributors*: J.M.Garg

**Image:Malva alcea pili NRM.jpg**  *Source*: http://en.wikipedia.org/w/index.php?title=File:Malva_alcea_pili_NRM.jpg  *License*: Creative Commons Attribution-Sharealike 3.0,2.5,2.0,1.0  *Contributors*: Aelwyn

**Image:Durio kutej F 070203 ime.jpg**  *Source*: http://en.wikipedia.org/w/index.php?title=File:Durio_kutej_F_070203_ime.jpg  *License*: Creative Commons Attribution-ShareAlike 3.0 Unported  *Contributors*: W.A. Djatmiko (Wie146)

**file:Theobroma grandiflorum-flower.jpg**  *Source*: http://en.wikipedia.org/w/index.php?title=File:Theobroma_grandiflorum-flower.jpg  *License*: Public Domain  *Contributors*: Puime

**file:Alcea_setosa.jpg**  *Source*: http://en.wikipedia.org/w/index.php?title=File:Alcea_setosa.jpg  *License*: Public Domain  *Contributors*: CarolSpears, Conscious, Matanya (usurped), Timichal, Werckmeister, Wst

**Image:2006 08 10 Hollyhock.JPG**  *Source*: http://en.wikipedia.org/w/index.php?title=File:2006_08_10_Hollyhock.JPG  *License*: Public Domain  *Contributors*: Original uploader was Indu Singh at en.wikipedia

**file:Euphorbia heterophylla (Painted Euphorbia) in Hyderabad, AP W IMG 9720.jpg**  *Source*: http://en.wikipedia.org/w/index.php?title=File:Euphorbia_heterophylla_(Painted_Euphorbia)_in_Hyderabad,_AP_W_IMG_9720.jpg  *License*: Creative Commons Attribution-Share Alike  *Contributors*: J.M.Garg

**file:Diversity of plants image version 5.png**  *Source*: http://en.wikipedia.org/w/index.php?title=File:Diversity_of_plants_image_version_5.png  *License*: Creative Commons Attribution-Share Alike  *Contributors*: Rkitko

**File:Haeckel Siphoneae.jpg**  *Source*: http://en.wikipedia.org/w/index.php?title=File:Haeckel_Siphoneae.jpg  *License*: Public Domain  *Contributors*: Dysmorodrepanis, Homonihilis, Ragesoss, Thiotrix, Vonvon, 2 anonymous edits

**File:Ferns02.jpg**  *Source*: http://en.wikipedia.org/w/index.php?title=File:Ferns02.jpg  *License*: unknown  *Contributors*: Amada44, Bdk, Fir0002

**File:Petrified forest log 1 md.jpg**  *Source*: http://en.wikipedia.org/w/index.php?title=File:Petrified_forest_log_1_md.jpg  *License*: Creative Commons Attribution-Sharealike 2.5  *Contributors*: User:Moondigger

**File:Leaf 1 web.jpg**  *Source*: http://en.wikipedia.org/w/index.php?title=File:Leaf_1_web.jpg  *License*: Public Domain  *Contributors*: Ies, Ranveig, Red devil 666, Rocket000, WeFt, Überraschungsbilder

**File:Eenbruinigherfstblad.jpg**  *Source*: http://en.wikipedia.org/w/index.php?title=File:Eenbruinigherfstblad.jpg  *License*: Public Domain  *Contributors*: Ischa1

**File:Dead plant in pots.jpg**  *Source*: http://en.wikipedia.org/w/index.php?title=File:Dead_plant_in_pots.jpg  *License*: Creative Commons Attribution-Sharealike 2.0  *Contributors*: vetcw3

**File:Plant cell structure svg.svg**  *Source*: http://en.wikipedia.org/w/index.php?title=File:Plant_cell_structure_svg.svg  *License*: Public Domain  *Contributors*: LadyofHats (Mariana Ruiz)

**File:VFT ne1.JPG**  *Source*: http://en.wikipedia.org/w/index.php?title=File:VFT_ne1.JPG  *License*: Creative Commons Attribution-Sharealike 2.5  *Contributors*: Aroche, BRUTE, ComputerHotline, Denis Barthel, NoahElhardt, 2 anonymous edits

**File:Potato plant.jpg**  *Source*: http://en.wikipedia.org/w/index.php?title=File:Potato_plant.jpg  *License*: Public Domain  *Contributors*: Scott Bauer

**File:Timber DonnellyMills2005 SeanMcClean.jpg**  *Source*: http://en.wikipedia.org/w/index.php?title=File:Timber_DonnellyMills2005_SeanMcClean.jpg  *License*: GNU Free Documentation License  *Contributors*: Original uploader was SeanMack at en.wikipedia

**File:Taxus wood.jpg**  *Source*: http://en.wikipedia.org/w/index.php?title=File:Taxus_wood.jpg  *License*: GNU Free Documentation License  *Contributors*: MPF

®
FSC
www.fsc.org
MIX
Papier aus verantwortungsvollen Quellen
Paper from responsible sources
FSC® C105338